Native Trees
and
Shrubs of the Florida Keys:
A Field Guide

J. PAUL SCURLOCK

Laurel Press, Inc.
Bethel Park, Pennsylvania

Copyright © 1987 by J. Paul Scurlock

All rights reserved. No part of this work may be reproduced or transmitted in any form by by any means, electronic or mechanical, including photocopying and recording, or by any information storage or retrieval system, without the prior, express, written permission of the author, excepting brief quotes used in connection with reviews written specifically for inclusion in a magazine or newspaper. No secondary edition of this work may be published without the express, written permission of the author or his agent.

Library of Congress Cataloging-in-Publication Data

Scurlock, J. Paul (James Paul), 1909-
  Native trees and shrubs of the Florida Keys : a field guide / J. Paul Scurlock. — 2nd ed.
      p.   cm.
  Includes bibliographical references and index.
  ISBN 0-9619155-2-8
  1. Trees—Florida—Florida Keys—Identification. 2. Shrubs—Florida—Florida Keys—identification. 3. Trees—Florida—Florida Keys—Pictorial works. 4. Shrubs—Florida—Florida Keys—Pictorial works. I. Title.
QK 154.S43 1992
582.1609759' 41—dc20   91-40892
                                                              CIP

Cover by Mary Craumer Scurlock

Edited by Mary-Alice Herbert

Available from Laurel Press, Inc.,
1514 Holly Hill Drive, Bethel Park, Pennsylvania 15102

To Mary

INTRODUCTION
PURPOSE OF THE BOOK

The purpose of this book is to provide an easy, positive means of identifying the trees and shrubs growing wild on the Florida Keys. More people come to the Keys every year; some make them their year-round home, and others spend from six months to only a few days here. Many who come are unfamiliar with the plants that grow in the tropical and subtropical climate of the Keys—it can be very frustrating to sail among and walk on these islands and not recognize and name the trees and shrubs. This book is a field guide to shrubs and trees native to the Keys.

There is a growing interest in native plants for the home garden, and local nurseries are beginning to stock them. A nursery established on Plantation Key, for instance, handles native plants exclusively. This is the result of an increasing demand for attractive plants tolerant of the ecological conditions of the area. People want plants that will thrive in spite of wet seasons, dry seasons, year-round insects, direct tropical sun, constant wind, salt spray and long periods of inattention when the homeowner is away. They are beginning to realize that native and naturalized plants are the answer, but they want to know what to choose. What plants have attractive flowers? Which are fragrant? What plants have edible fruit? Which provide food for birds? How large does a certain plant grow? This book is a guide for gardeners in choosing plant material ideally suited to growing conditions in the Keys.

According to Elbert L. Little in the *Atlas of United States Trees, Volume 5, Florida,* "the Keys have about 102 species of native trees, 13 of which, being absent from the mainland, are found only in the Florida Keys." Scurlock located and photographed approximately 160 plants most authorities consider "native"—34 are listed by the Florida Game and Fresh Water Commission in its *Endangered and Potentially Endangered Fauna and Flora in Florida* (1 October 1983). For naturalists, botanists, conservationists, and the increasingly interested public, Scurlock catalogs the Keys' valuable and irreplaceable botanical heritage; it is hoped, this book will encourage continued interest and an even greater effort to save some of the endangered species.

SCOPE OF THE BOOK

The plants photographed and described here are those found on "The Florida Keys," the chain of islands along Route U.S. #1, from Key Largo to Key West, including islands off-route or accessible by boat, such as Lignum Vitae, No Name, Big Torch and Johnston keys, and the islands in Biscayne National Park. Over ten years, the author attempted to photograph specimens of *all* native trees and shrubs in this area. *These plants, however, are also recognized as natives of South Florida, Cuba, the Bahamas, the islands of the Caribbean, parts of Mexico, South and Central America.*

In addition to those plants generally considered to be native to the Keys, several introduced species have become so well-adjusted to the Key's environment that they can be found growing wild in large numbers. Since the people using this book will invariably find them on their field trips, these plants have been included. Some examples are: Australian Pine *(Casuarina equisetifolia)*, Asiatic Colubrina *(Colubrina asiatica)*, Guava *(Psidium gaujava)*, Brazilian Pepper *(Schinus terebinthifolia)*, and the Seaside Mahoe *(Thespesia populnea)*.

Some of the native plants could not be found growing in wild areas under conditions satisfactory for photography—for example, those in the heart of dense, dark hammocks. If such a plant was available in a botanical garden, municipal park, or private garden, it was photographed there—only as a last resort to illustrate the species. *In every case the plant was growing in the Keys, and it was photographed by the author personally.* The descriptive notes indicate the Key on which it was photographed and, if it was in flower, the month is also given. Beekeepers, for instance, are interested in the flowering months of plants, so as to schedule the placement of hives in all seasons. Winter residents, visitors, and gardeners are also interested in the blooming season.

THE FORMAT

The plants illustrated and described in this book are arranged in alphabetical order by botanical name. Since common names vary from one locality to another, the arrangement by scientific name should result in less confusion. With the help of the well-known taxonomist, George N. Avery, and the University of Florida Institute of Food and Agricultural Sciences Plant Identification and Information Service, the latest approved scientific name for each plant has been used. To make it easier for a person trying to identify a plant, the Index and text include some well-known common names and scientific synonyms.

Whenever possible, photographs were chosen to show those characteristics which best identify a plant—its bark, fruit, leaves, spines, height, shape, and so forth. In the field it is difficult to find the plants in flower or fruit at different times of the year—using the text to supplement the pictures should ease identification. Realizing that people using this book may be unfamiliar with some of the technical terms that usually appear in manuals, the author used everyday language as much as possible, except where technical terms were necessary to conserve space. A brief glossary of technical terms is included in page 173.

The use of the scientific names of the plants in alphabetical order at times separates some species from their natural family groups or from groups having similar characteristics. To offset this, an index of the plant families and the different species included is shown in an Index to Family Groups on page 202.

An Identification Key to link characteristics of a plant in the field to its picture and description begins on page 178. Identification keys can be so complicated and technical that they discourage all but the professional botanist. This one has been kept as simple as possible: in addition to leaf form (simple, compound, alternate, opposite, length and shape), significant characteristics such as color and size of flowers and fruit, and the presence of spines are listed. These characteristics offer clues to identity throughout the year: leaves are usually present throughout the year; eye-catching flowers, fruits, or both are often present on most plants. Because certain groups are easily recognized by their general character—Palms, Pines, Cacti, Agave—these are listed first.

Many of us involved in the use, administration, control, and protection of native plants, need not only to identify them but also to know their status. Homeowners, for example, can take particular pleasure in and make a major contribution towards conserving protected plants by developing their gardens with as many native plants as possible, especially those considered *endangered, threatened, rare,* or of *special concern*. Enlightened developers, road-builders, conservationists, naturalists, educators, and government personnel are among those who also need to know. Plants listed by the Florida Game and Fresh Water Commission as having special status are so noted on the individual plant pages. All special status plants are listed and the status designations defined on page 209.

ACKNOWLEDGMENTS

This book is the result of about ten years of wading, crawling, and tramping in the Florida Keys. It would have been impossible without some special people, a few of whom I can mention here:

George Avery, taxonomist, who helped locate some of the rare specimens.

Janet Bunch, Horticulturist, Florida Cooperative Extension Service.

Jeffrey Fisher, Director, Monroe County Extension Service, who encouraged me to complete this study.

David Hall, Extension Botanist, Herbarium, University of Florida, who assisted with the design of this book.

Lois Kitching, Director, Key West Garden Club.

Jeanne Parks, Ranger, Lignum Vitae Key.

John Popenoe, Director, Fairchild Tropical Garden.

Alexander Sprunt, Field Research Director, National Audubon Society.

Donna Sprunt, operator of the Florida Keys Native Nursery, Inc., Plantation Key.

Arthur Weiner, Project Director of the Ecological Study of the Hardwood Hammocks of the Florida Keys.

Ann Williams, President, Big Pine Key Botanical Society.

Kathy Wolf, City of Key West Landscape Coordinator and Instructor of Botany, the Florida Keys Community College.

Ray Zerba, Florida Cooperative Extension Agent, Monroe County, a friend who encouraged and pressured me to publish this manuscript.

CONTENTS

Introduction v
Native Trees and Shrubs of the Florida Keys 1
Selected References 171
Glossary 173
Identification Keys 178
Identification Keys Index 179
Index to Family Groups 202
Index to Special Groups 207
Designated Status 209
Index to Scientific and Common Names 210
About the Author 219

Abutilon permolle (Willd.) Sweet   MALVACEAE
Common Name: **Indian Mallow.**

This shrub is native to South Florida, Mexico, Central America, the West Indies and the Florida Keys. It normally reaches a height of 4 to 5'. Its simple, heart-shaped leaves are mostly 2 to 5" long. Its stems are densely pubescent. The leaves are deeply notched at the base where the stems are attached and have irregularly scalloped margins.

The bright yellow flowers appear solitary on stalks ½ to 1½" long in the leaf axils. The 5 petals are spread out in a round, flat plane similar to the Rose Mallow. The corolla is 1 to 1½" across.

The fruit or seed pod is composed of 7 to 10 radial sections which open when ripe to spill out many small seeds.

The plants shown here were photographed on Plantation and Big Pine Keys in May.

Acacia choriophylla Benth. FABACEAE
Common Names: **Cinnecord, Tamarindillo, Frijollo.**

A very rare tree on the Florida Keys, it is also native to the Bahamas and Cuba. Unlike many of the *Acacias*, it has no spines along the stems. It does, however, have small stipular spines in the leaf axils. The leaves are alternate, bipinnate and have 1 to 3 pairs of pinnae. The pinnae are opposite and have 3 to 5 opposite pairs of leaflets which are dark green above, paler below with entire margins turned under slightly. They are leathery, elliptic or oblong in shape, rounded or slightly notched at the apex and from ⅓ to 1⅓″ long.

The flowers are bright yellow in spherical heads about ⅜″ in diameter and clustered in panicles from the leaf axils. The fruits are flat, garden-pea shaped pods about 2″ long containing 2 to 9 seeds. The seeds sprout readily, providing an easy method of propagation. The tree requires very little attention after it gets started. It can be trimmed to any desired shape.

The specimen shown here was photographed in April on Sugarloaf Key.
DESIGNATED STATUS: Endangered.

Acacia farnesiana (L.) Willd.   FABACEAE
Common Names: **Sweet Acacia, Opopanax, Popinac, Cassie, Huisache.**

The sweet acacia grows either as a many-stemmed shrub in clumps or as a small tree to 10 or 12' tall. Its feathery and fern-like compound leaves are bright green with small leaflets about ¼" long. It is deciduous but the leaves are persistent except when the cold North wind blows. There are 1 to 1½" sharp spines along the branches. The flowers are small, bright yellow, very fragrant and occur in globular clusters ½ to 1" in diameter on short stems growing out of the leaf axils. They bloom intermitantly almost all year. The fruits are dark-brown, cylindrical pods about 3" long filled with pulp and flat, brown, polished oval seeds.

Legend claims perfume for Cleopatra (Queen of the Nile) was made from the flowers. It is cultivated as a source of perfume in Europe.

It is native to Florida, Mexico, the West Indies and the Florida Keys. The above photographs were made on Sugarloaf Key—the flowers in December and the fruit in March.

SCIENTIFIC SYNONYM: *Vachellia farnesiana* (L.) Wight & Arn.

Acacia macracantha Humb. & Bonpl. ex Willd.   FABACEAE
Common Name: **Long-spine Acacia.**

This small tree is found in the hammocks of the West Indies, Central and South America and the Florida Keys. It is very rare in the Keys.

The branches are armed with hard-paired spines up to 1½" and more in length. They are persistant on the main trunk. The leaves are bipinnate, about 4" long with 8 or more pairs of pinnae. The many leaflets are obtuse-oblong in shape, about ⅛" long and very narrow with prominent mid-veins. The fruits are bean-like pods about 3" long when fully mature and are laterally compressed.

Twenty plants of varying heights up to 20' were found on a limestone sandy ridge on Ramrod Key. The flowers in globular clustered heads about ¼" across were photographed in November.

SCIENTIFIC SYNONYM: *Poponax macracantha* Humb. & Bonpl. ex Willd.

Acacia pinetorum Hermann — FABACEAE
Common Name: **Pine Acacia.**

This shrub or small tree ranges from Lee County, Florida south through the Keys and the West Indies. The leaves are compound with 3 or 4 pairs of pinnae. The shrub is very similar to the *A. farnesiana*. Its leaflets, however, are only about ½ the size of those of the *A. farnesiana*. The seed pods are pointed or beaked instead of blunt like those of the *A. farnesiana*. The pale gray, glabrous spines are up to 1" long.

It is found growing usually on the edges of hardwood hammocks. The leaves give it a very attractive feathery look and serve as a backdrop for the golden globular heads of the fragrant flowers.

The above plants were photographed on Big Pine Key in June.

SCIENTIFIC SYNONYM: *Vachellia peninsularis* Small.

Acoelorrhaphe wrightii (Griseb. & H. Wendl.)   ARECACEAE
Wendl. ex Becc.

Common Names: **Paurotis Palm, Everglades Cabbage Palm, Saw Cabbage Palm, Madeira Palm, Silver-saw Palmetto.**

This is a cluster-forming palm of many slender matt-covered trunks, at times making very large clumps with some trunks reaching a height of 20 to 30'. The leaves are fan-shaped, medium green on both sides, about 2' in diameter and divided only to the middle of the leaf. The leaf stems are thin, 2 to 3' long with orange-colored saw teeth on the edges.

The small, yellow-green flowers grow on a stalk coming from among the leaves and extending beyond them. The fruit is globular in shape and orange-colored turning black as it ripens forming a smooth, hard, round seed.

It is native to South Florida and the Everglades, the Bahamas and the West Indies. It is not currently known to exist in the wild state in the Keys.

It can be propagated from seed or division of clump. It requires more water than average for good growth.

It was photographed in a private garden on Summerland Key in December.

SCIENTIFIC SYNONYM: *Paurotis wrightii* (Griseb. & Wendl.) Britt.
DESIGNATED STATUS: Threatened.

Agave decipiens Baker   AGAVACEAE
Common Names: **False Sisal, Century Plant.**

This plant grows in the wild state throughout the Florida Keys. Some authorities believe, however, it was introduced prehistorically from Mexico. It normally has a trunk from 2 to 8' high when mature.

The leaves are lanceolate in shape, deeply concave, up to 4" wide, 3' or more long, green and succulent. They are outcurving with sharp, recurved spines in the margins. They terminate in a hard, brown, needle-like point. The greenish-yellow flowers grow on a branched scape that may be 25' high. The fruit capsules are ellipsoid in shape, 1 to 1½" long with many thin flat seeds. Usually many bulbils or aerial-bulbs form on the scape, rooting in clumps before falling to the ground. The individual plants flower only once and then wither and die. They produce suckers or offsets at the base that mature later.

The plants shown here were photographed on Windley and Big Pine Keys in August and September.

Agave sisalana Perrine　　　　　　　　　　AGAVACEAE
Common Names: **Century Plant, Sisal Hemp, Sisal, Sisal Agave.**

This plant was introduced into the Keys from the Yucatan of Mexico in 1836 by Dr. Henry Perrine. Indian Key, where Dr. Perrine was killed in the Indian Massacre of 1840, has a very large colony of the plants today. They are growing wild in many of the Keys. It has 4-inch wide, sword-like, fibrous leaves that extend upward 5 or more feet from a rosette at the base. They terminate in long, hard, sharp spines. They are green or grayish in color, smooth in texture and contain a liquid that may be a skin irritant for some people.

　The flower stalk may rise 25 to 30' forming a much branched top on which many greenish flowers about 2½" wide appear. The fruit is an egg-sized capsule which splits open releasing black seeds. Germinating seedlings or bulbils develop along the flower stalk. After blooming once, the mother plant dies. Root suckers sprout around the base of the plant.

　The plants were photographed on Plantation Key and Sugarloaf Key in August.

Albizia lebbeck Benth.   FABACEAE

Common Names: **Woman's Tongue, Mother-in-Law's Tongue, Siris Tree, Lebbeck Tree.**

This is a very large tree, reaching a height of 50' or more in the Keys. The width of the crown will usually match the height of the tree. It is a native of tropical Asia and Northern Australia but has become naturalized in the Keys.

The leaves are pinnately compound, deciduous and usually have up to 10 opposite pairs of medium-green leaflets. The leaflets are ovate in shape and about 2½" long. The flowers which appear generally throughout the year are small, light-yellow and occur in fluffy globular clusters. The seed pods are flat and about 8" long. When ripe, they become tan in color, very dry and contain 4 or 5 hard, loose seeds. They are persistant and rattle in the wind—thus the popular name "Woman's Tongue."

The dark-brown wood is used for making furniture. The seeds, bark and flowers are said to be used in India for treatment of leprosy, skin diseases, mumps, eye ailments and other problems.

The flower pictures were taken in Key West and on Sugarloaf Key in April.

SCIENTIFIC SYNONYM: *Mimosa lebbeck* L.

Amphitecna latifolia (Mill.) A.H.Gentry   BIGNONIACEAE
Common Name: **Black Calabash.**

The black calabash is native to the West Indies, Mexico, Central America and South Florida. Although some authors have reported it as being native to the Florida Keys, this author has found none growing in the wild to date. It is an upright tree with many vertical branches making a very dense, oval-shaped crown. The leaves are evergreen, oval-shaped, simple, alternate, entire and 6 to 8" long. They are dark-green, glossy and short-pointed or rounded at the apex.

The flowers are creamy-white to purplish. They are bell-shaped, about 2½" long and almost hidden by the thick leaves. They usually bloom twice annually and are not fragrant. The fruit is subglobose or melon-shaped, 3 to 4" long and 2 to 3" wide. It is not edible but when dried makes a good drinking cup.

It was photographed in a private garden on Sugarloaf Key in flower in December.

SCIENTIFIC SYNONYM: *Enallagma latifolia* (Mill.) Small.

Amyris elemifera L. RUTACEAE
Common Name: **Torchwood.**

This plant is native to South Florida, the Bahamas, the West Indies, Central America and the Florida Keys. It is a medium size tree or shrub with light brown bark and slender, unarmed leaf-scarred stems. It is very resinous and oils have been extracted for certain pharmaceutical purposes. The wood is used for fuel and torches. When in bloom, the plant is covered with very fragrant flowers.

The compound leaves are light green and have 3 to 5 drooping, opposite leaflets that are ovate to ovate-lanceolate with long-tapered apexes. They are 1 to 2½" long with entire or finely scalloped margins.

The flowers have 4 white petals, 8 stamens and occur in terminal clusters. The fruits are purple or black, globular or ovoid drupes about ½" long containing a single seed and thin, aromatic black pulp, supposedly edible.

The plant was photographed on Cudjoe Key in December.

SCIENTIFIC SYNONYM: *A. maritima* Jacq.

Annona glabra L. — ANNONACEAE

Common Names: **Pond Apple, Alligator-apple, Corkwood, Custard-apple.**

This tree is native to South Florida, the West Indies, the Bahamas and the Florida Keys. It usually reaches a height of 20' or less in the Keys. It has simple, smooth, leathery leaves from 3 to 5" long with entire margins. They are bright green, with yellowish veins, oblong to elliptical in shape and acute at the apex. They are deciduous but usually remain on the tree until the new growth appears.

The flowers are solitary with 2 three-petal whorls of yellowish-white petals. They are triangular in shape and the outer set of petals have a red spot near the base. The flowers are about 1" wide. The compound fruit is usually solitary, in the shape of an inverted pear, 3 to 5" long and ripening to a yellowish color with brown blotches. It is edible but not tasty.

The wood is very light and has been used for fishing floats. The tree is usually found growing in swamps, shallow ponds and sink holes.

This specimen was photographed on Big Pine Key in July.

SCIENTIFIC SYNONYM: *A. palustris* L.

Ardisia escallonioides Schiede & Deppe ex Schlecktend. & Cham.   MYRSINACEAE

Common Names: **Marlberry, Marbleberry, Cherry.**

A shrub or slender tree to 20', it has light grayish bark and a narrow columnar crown. It is native to Mexico, the Bahamas, the West Indies and the Florida Keys.

The leaves are evergreen, lance shaped, 4 to 6" long, glossy and alternate. The flowers are white and purple with yellow stamens, fragrant and about ¼" wide in large terminal clusters. The red fruits, turning dark purple when ripe, are glossy, 1-seeded drupes about ¼" in diameter. They are edible but dry and too acid to be desirable. It is reported that the Mikasuki Indians mix the leaves with their smoking tobacco.

The fruit shown here was photographed on Sugarloaf Key in December. The flowers were photographed on Long Key in October.

SCIENTIFIC SYNONYM: *Icacorea paniculata* (Nutt) Ludw.

Argythamnia blodgettii (Torr.) Chapm.　　EUPHORBIACEAE
Common Names: **Argythamnia, Blodgett's Wild-mercury.**

This shrub grows along the edges of hammocks in low, moist soil in South Florida and the Florida Keys. It has erect stems that reach a height of 2' or more. The leaves are ovate to elliptic or spatulate in shape and 1 to 2" long. They are alternate, pubescent and have short petioles.

The plant is monoecious with both male and female flowers in the same raceme. The staminate flowers are usually above the pistillate flowers. Both kinds have 5 sepals and 5 petals and are about ¼" wide.

The fruit is a 3-lobed capsule less than ¼" wide and contains subglobose seeds.

The plant was photographed on Plantation Key in August.

SCIENTIFIC SYNONYM: ***Ditaxis blodgettii*** Vahl.

DESIGNATED STATUS: Threatened.

Ateramnus lucidus (Sw.) Rothm.   EUPHORBIACEAE
Common Names: **Crabwood, Oysterwood.**

This shrub or small tree 20 to 30' high is found in hammocks and on wooded shores of South Florida and the Keys, the Yucatan and the West Indies. The sapwood is very light yellow contrasting sharply with the dark heartwood. It is a favorite for fancy turning because of its high contrast between the heart and sapwood. The leaves are dark green in color with rough veins on the upper surface, 2 to 4" long and ½ to 1½" wide. They are alternate, leathery, persistent and elliptical in shape with smooth or wavy margins sometimes toothed near the apex.

Both male and female flowers occur on the same tree: the male are very small, yellow-green and numerous on spikes 1 to 2" long; the female, single on a longer stalk. Both are axillary and fragrant. The fruits are globular capsules, dark reddish-brown or nearly black in color. They are ¼ to ½" in diameter with thin, dry flesh covering globular seeds.

The illustrative specimen was photographed in flower on No Name Key in May.

SCIENTIFIC SYNONYM: *Gymnanthes lucida* Sw.

Avicennia germinans (L.) L.   AVICENNIACEAE
Common Names: **Black Mangrove, Honey Mangrove, Saltbush.**

Along with the red and white mangroves, this is one of the most important plants in the Keys. It grows in the muddy saline area, sending up masses of breather roots (pneumatophores) that collect debris and silt, building up the shoreline and preventing erosion. Of all of the mangroves, it is the most tolerant of high salinity. It is native to Florida, the West Indies, most of Tropical America and the Florida Keys.

It occurs as a shrub or a tree to 25 or 30' tall. Its leaves are opposite, elliptic, 2 to 5" long, green above and gray and salty below. The salt can be collected from the leaves for seasoning.

The pale-yellow or white flowers are tubular, 4-lobed, and about ½" wide in terminal or axillary spikes. They are fragrant and bloom throughout the year but most conspicuously in June and July. They are the main source for the well-known "mangrove honey." The fruit is an egg-shaped pod about 1½" wide, flattened on two sides and containing 1 seed. Frequently some of the seeds sprout on the tree.

The tree in flower was photographed on Sugarloaf Key in May.

SCIENTIFIC SYNONYMS: *A. nitida* Jacq., ***Bontia germinans*** L.
DESIGNATED STATUS: Special Concern.

Baccharis angustifolia Machx.   ASTERACEAE
Common Names: **False Willow, Saltbush.**

Like the *B. halimifolia*, this plant is found in the coastal region of Southeastern U.S. from North Carolina to Texas, Florida, the West Indies and the Florida Keys. It has many erect branches and may reach a height of 10' or more in the Keys.

The leaves are narrowly linear to linear-lanceolate with entire or toothed edges. They are alternate and vary in length from 1 to 3½". Male and female flowers appear on separate plants. Since the plants are not fragrant and have no nectar to attract insects, pollination is accomplished by the wind. The flowers are whitish, greenish or yellowish, very small and occur in compound heads. The female flowers develop ribbed achenes equipped with hair-like bristles that act as sails in the distribution of the released seeds (lower left picture). The male flowers are on right.

These plants were photographed on Upper Key Largo in October.

Baccharis halimifolia L. ASTERACEAE

Common Names: **Saltbush, Sea Myrtle, Silverling, Groundsel Tree, Eastern Baccharis.**

This shrub or small tree is native to Florida, the West Indies, the coastal areas of Eastern U.S.A. and the Gulf of Mexico and the Florida Keys. It grows in salt marshes and low-ground hammocks and may reach a height of 10' or more.

The leaves are evergreen, alternate, spatulate in shape, light green, deeply toothed or notched and 1 to 3" long. The flowers are small, white or greenish-yellow, tubular, in compound heads on axillary stalks. Male and female flowers are on different plants. White hair-like bristles extend beyond the leafy bracts of the female flowers giving the plant a silvery appearance and aiding in the dispersal of the matured small seeds.

The above pictures were made on Sugarloaf Key in October and November.

Borrichia arborescens (L.) DC. ASTERACEAE
Common Name: **Sea Ox-eye.**

This plant is found in the tidewater areas, along the beaches and in the low hammocks. It is a shrub, at times reaching a height of 3 to 4'. As the species name *"arborescens"* implies, it is frequently found in a small tree-like form. The leaves are light-green in color, 1 to 3" long, spatulate or oblanceolate in shape with a small pointed tip at the apex. They are fleshy, opposite and have faintly defined veins running parallel to the long-side margins.

The flowers are composite heads up to 1" across with bright yellow rays and bloom off and on all year. They appear singly at branch terminals and look like small sunflowers. The phyllaries or outer bracts of the flower heads are *not* tipped with sharp spines. This feature is the principal difference between this species and the "Sea Daisy" *(B. frutescens)* which does have spines on the bracts.

The plant is native to the Keys, South Florida, Tropical America and the West Indies.

The specimen pictured here was photographed on Sugarloaf Key in May.

Borrichia frutescens (L.) DC.   ASTERACEAE
Common Name: **Sea Daisy.**

The Sea Daisy is very similar to the Sea Ox-eye *(B. arborescens)*. It is also found along the shorelines, in the salt marshes and along roadsides and trails. In fact, very often it can be seen growing in the same areas throughout the Keys. It ranges through Florida north to Virginia and west into Texas. The opposite, thick, grayish-green, 1 to 3″ long leaves are narrow and broadened at the apex with a small projection at the tip. They are succulent, faintly aromatic, covered with very fine gray hair and sometimes have shallow toothed edges.

The composite-headed flowers with yellow petals are ½ to 1″ across, grow on terminal branch tips and appear all year in the Keys. The phyllaries or leaf-like bracts of the flowers are tipped with sharp spines and curve outward and downward and are the main features that distinguish the Sea Daisy from the Sea Ox-eye *(B. arborescens)*.

The specimen shown here was photographed on Geiger Key in February.

Bourreria cassinifolia (A. Rich.) Griseb.   BORAGINACEAE
Common Names: **Smooth Strongbark, Little Strongback.**

This shrub or small tree is reputedly native to the pinelands of Florida, the West Indies and the Florida Keys. With the help of George Avery, the only known stand of this plant in the Keys was found on Big Pine Key. The tallest plant was about 7'.

The leaves are elliptic or ovate in shape and from ½ to ¾" long. They have accented veins. The flowers are very similar to the ***B. ovata*** and the ***B. radula***. They are white, about ½" across, bell shaped with 5 lobes and were found as solitary blooms.

The fruits are depressed globular drupes about ¼" in diameter and yellow-orange in color.

The plants were photographed in flower in November.

Bourreria ovata Miers BORAGINACEAE

Common Names: **Strongbark, Bahama Strongbark, Oval-leaf Strongbark, Strongback.**

This plant is a shrubby tree, native to South Florida, the Bahamas, the West Indies and the Florida Keys. It may reach a height of 15 to 30'. It has reddish-brown bark from which the Bahamians are said to make tea.

The leaves are persistent, 2 to 3" long, oval in shape with notched or rounded tips. They are green with orange tinted midribs, glossy above, pale below on long drooping twigs.

The flowers are creamy white, about ½" across, bell shaped with 5 lobes and grow in terminal clusters. The fruits are globular drupes about ½" in diameter. They change from green to yellow to orange to red as they ripen—frequently all shades occur in one cluster. They are edible but not very tasty.

The plant is very attractive to butterflies and birds.

The pictures were taken in Key West in October and Plantation Key in August.

Bourreria radula (Poir.) G. Dou   BORAGINACEAE
Common Names: **Rough Strongbark, Rough Strongback.**

This shrub or small tree is considered to be native to South Florida, the West Indies and the Florida Keys. It is now quite rare. The *Atlas of United States Trees,* Volume 5, *Florida* by Elbert L. Little, Jr. shows it to exist in Key West and the adjacent Lower Keys only.

It is very similar to the **Bourreria ovata** except the leaves are smaller and densely hairy and the fruit is larger.

The above pictures were taken in the residential section of Key West. The plant was probably cultivated. It was photographed in June and October.

SCIENTIFIC SYNONYMS: *B. succulenta* Jacq., *B. revoluta* H.B.K.

Bumelia celastrina HBK. SAPOTACEAE
Common Names: **Saffron Plum, Downward Plum, Antswood.**

This shrub or small tree is native to Florida, the West Indies and the Florida Keys. It has black furrowed bark, a dense crown and spine-tipped twigs. The wood is very hard, heavy and strong. The largest one recorded in the U.S.A. is on Big Pine Key. It is 23' high, has a 17' crown spread and is 27" in circumference.

The small light-green leaves are alternate, sometimes clustered along the branches, blunt-pointed, ½ to 1½" long and new growth is frequently pinkish in color. The flowers are light green, inconspicuous and fragrant. The fruits are green, oblong-globular drupes turning black when ripe. They are ½ to 1" long, sweet, edible raw but quite sticky.

The photographs shown here were taken on Stock Island and Sugarloaf Key in April.

SCIENTIFIC SYNONYM: ***B. angustifolia*** Nutt.

Bumelia salicifolia (L.) Sw.   SAPOTACEAE
Common Names: **Willow Bustic, Bustic, Cassada.**

This small evergreen tree whose upright branches form a slender symmetrical crown may reach a height of about 30'. It is native to South Florida, the West Indies and the Florida Keys. New growth is covered with a fine hairy fuzz. The older bark is ashy-gray or light brown.

The leaves are dark green above and pale below, curving, lanceolate in shape, alternate, pointed, persistent and from 3 to 5" long, more or less crowded at the end of the twigs.

Small white 5-petaled flowers are borne in thick clusters in the leaf axils and along the branches for a distance of 8 to 12". The fruit, single or in clusters, is a black berry when ripe. It is oblong to globular in shape, about ¼" long and contains 1 or more pale brown seeds. The wood is very hard, heavy, close-grained and dark brown or red in color.

It was photographed in flower on Sugarloaf Key in May.

SCIENTIFIC SYNONYM: *Dipholis salicifolia* (L.) A.DC.

Bursera simaruba (L.) Sarg.　　　　BURSERACEAE
Common Names: **Gumbo-limbo, West Indian Birch, Gum-elemi, Tourist Tree.**

One of the most popular trees in the Keys is the Gumbo-limbo. It is native to South Florida, the West Indies and the Florida Keys. It is a large shade tree reaching a height of 40' or more with an equally wide crown. Its trunk is large with reddish-brown, peeling, papery bark.

The leaves are deciduous, alternate and pinnately compound with 3 to 9 glossy, green, ovate to oblong, pointed leaflets 2 to 3" long. The flowers are small and greenish-white in many-flowered panicles, followed by dark red fruits containing 1 or 2 hard-shell seeds. The tree is easily propagated by placing cuttings directly in the soil. Living fences were observed in Costa Rica along highways made from cuttings of the tree.

The aromatic resin reportedly was used in the treatment of gout and in the manufacture of varnish. The leaves have been used in making tea.

The tree was photographed in flower on Stock Island in May.

SCIENTIFIC SYNONYM: *Elaphrium simaruba* (L.) Rose.

Byrsonima lucida (Mill.) DC. MALPIGHIACEAE
Common Names: **Locust-berry, Key Byrsonima.**

A native of the Florida Keys, the Everglades and the West Indies, it can be a many-stemmed shrub or a small tree. The above specimen was photographed on Sugarloaf Key and is 15' high with a 28' crown. The bark is light brown and smooth. It has spreading branches and a wide flat-topped head, usually found in the hammocks and pinelands.

The leaves are opposite, dark green and shiny above, dull and paler below, obovate or spatulate in shape and 1 to 2½" long. They are usually round-tipped but may have a small point.

The flowers are shite or pink changing to yellow or rose-red. They have 5 petals, are about ¼" wide and grow in clusters 1½" across. The fruit is greenish turning brown when ripe. It is globular in shape, ¼" in diameter and covers a hard stone that contains a seed. The fruit is edible but not very tasty.

The flower pictures were made in April.

SCIENTIFIC SYNONYM: *B. cuneata* (Turcz) P. Wilson.

Caesalpinia bonduc (L.) Roxb. FABACEAE

Common Names: **Gray Nicker Bean, Sea Bean, Fever Nut, Hold-back, Wait-a-bit Vine.**

This erect or reclining, prickly shrub is native to South Florida and the Keys, the Bahamas and the West Indies. The vine-like branches, covered with curved spines, climb over other vegetation, frequently reaching a length of 15 to 20'. The bipinnate leaves with 4 or more pairs of pinnae are 15" or more long. The leaflets are ovate and 1 to 2" long.

The flowers are dull yellow, about ½" across, with 5 petals in axillary clusters up to 12" long. The fruits are oval-elliptic, 2 to 3" long, spiny, reddish-brown pods. When ripe and dry the pods split open upward to expose 1 or 2 gray, oval, smooth, hard seeds ¾" in diameter. These are the most commonly found floating "Sea Beans." Natives of the West Indies carry them as "Worry Stones" or "Pet Rocks." The kernels are reportedly used as a substitution for quinine and the new leaves are used for toothache in Ceylon. The hard seeds are often strung in necklaces.

The above plant was photographed on Geiger Key in bloom in November.

SCIENTIFIC SYNONYMS: *C. crista* L., *Guilandina bonduc* L., *Guilandina crista* (L.) Small.

Caesalpinia major (Medic.) Dandy & Exell   FABACEAE
Common Names: **Yellow Nicker, Yellow Nicker-bean.**

This sprawling, woody shrub is very similar to the *C. bonduc* except its leaflets are relatively larger and its bean-like seeds are yellow instead of gray.

Climbing on itself and other nearby plants, it frequently forms a thorny, impenetrable mass. It is found in the West Indies, Florida and the Florida Keys.

The plant with fruit was photographed on the edge of a hammock on Lower Matecumbe Key in October. The flower picture was taken on Big Coppitt Key in November.

SCIENTIFIC SYNONYMS: *C. bonduc* Roxb., ***Guilandina bonduc*** L.

Caesalpinia pauciflora (Griseb.)
C. Wright ex Sauvella

FABACEAE

Common Name: **Caesalpinia.**

This shrub is erect reaching a height of 6' or more. It has compound bipinnate leaves with 4 or more pairs of pinnae having 10 to 14 glabrous leaflets. The leaflets are oblong, oblique, slightly notched at the apex and about ½" long. There are paired, stiff spines about ¼" long at the leaf bases.

The yellow flowers are followed by a pea-shaped pod containing 2 or more seeds. It is tapered at each end and from 1 to 1½" long.

The plants were photographed in fruit on Big Pine Key in October. The flower picture was taken in February.

Callicarpa americana L.   VERBENACEAE
Common Names: **Beautyberry, American Beautyberry, French Mulberry.**

This shrub is native to South Florida and the Florida Keys. It is usually found in the hammocks and pinelands. It is a compact shrub with slender, outspreading stems growing from 3 to 7' high. The light green leaves are oval in shape, pointed at both ends, rough in texture, opposite, have finely toothed edges and are deciduous.

The white or pale pink flowers are clustered in the leaf axils. They are small, 4-petaled and about ¼" long. The ripe fruits are pink, violet or (rarely) white ⅛" wide globular drupes massed along the stems. They are edible but not very tasty and are a choice food of mockingbirds.

Flowering plants were photographed in June and September on Plantation Key and Islamorada. The fruiting plant was photographed in Key West in January.

Calyptranthes pallens Griseb. MYRTACEAE
Common Names: **Pale Lidflower, Spicewood.**

This shrub or small tree seldom grows above 20' in the Keys. It is also found in South Florida, the Bahamas, the West Indies, Mexico and Guatemala. The leaves are opposite, leathery, elliptical in shape, long tapered at the apex and base, up to 3" long and have entire margins. The young leaves unfold pink or light red in color, turning dark green when mature. They are aromatic—thus the common name "Spicewood."

The small, ⅛" long, greenish flowers covered with very fine hair, borne in long, stalked axillary clusters about 2½" long have no petals but many stamens. The buds are enclosed by an almost circular, lid-like limb which opens as the flower emerges. The fruit is a globular berry about ¼" in diameter with juicy, edible pulp covering 2 or 4 seeds. The berries finally turn purplish-black when ripe.

The flower pictures were taken on Sugarloaf Key in June.

Calyptranthes zuzygium (L.) Sw. MYRTACEAE
Common Names: **Myrtle-of-the-River, Spicewood.**

This tree is native to South Florida, Tropical America, the Bahamas, the West Indies and the Florida Keys. It is shrubby with light gray bark, a regular system of forking and may reach a height of 20' or more in the Keys.

The leaves are flat, dull, dark yellow-green and almost stemless, differing from the *C. pallens* whose leaves have longer stems. The leaves are opposite, elliptic-ovate in general shape, tapering to a point at the apex, slightly wavy margins, 1 to 3" long, 1 to 1½" wide and have slightly raised center veins.

The flowers have a greenish-white calyx, many stamens and no petals. They are stalked, about ¼" wide, solitary and occur in a branched inflorescence which is shorter than those of the *C. pallens*. The fruit is a depressed globular, 1 or 2-seeded berry, about ⅓" in diameter, which turns red and finally purplish-black as it ripens, retaining its hypanthium rim.

The plant was photographed on Key Largo in March.

Canella winterana (L.) Gaertn.   CANELLACEAE
Common Names: **Cinnamon Bark, Wild Cinnamon.**

This very attractive hardwood tree found throughout the Keys reaches a height of 30' or more with a straight trunk up to 10" in diameter. The inner bark and leaves are pleasantly aromatic, the pale yellow inner bark being the "Wild Cinnamon Bark" of commerce. The bright-green, leathery leaves are obovate, have a rounded or shallow notched apex, a narrow wedge-shaped base, and are 3½ to 5" long and 1½ to 2" wide. They are alternate, persistant and have entire margins with short, stout and grooved petioles.

The flowers have 5 white to red to purple petals. They are about ⅛" in diameter and appear in terminal or subterminal clusters. The fruit is a soft and fleshy, bright red, 2 to 4-seeded berry, about ½" in diameter and tipped with a persistant style. The seeds are less than ¼" long and are black and shiny.

The tree was photographed in flower on Big Pine Key in June and with fruit on Plantation Key in February.

SCIENTIFIC SYNONYM: *C. alba* Murray.

Capparis cynophallophora L.   CAPPARACEAE
Common Names: **Jamaica Caper, Caper Tree, Black Willow.**

The Jamaica Caper is an evergreen shrub or small tree with reddish-brown bark, at times reaching a height of 20' or more. It is native to South Florida, the Keys, the West Indies and Tropical America. The leaves are light-green above, leathery, oval in shape and rounded or notched at the apex. They have rusty scales on the underneath side, are 2 to 3" long and have smooth, entire margins that are slightly turned under.

The fragrant flowers have 4 white petals and many 1 to 2" long, purple filaments with yellow anthers extending beyond the petals in a brush-like form. The fruit is a slender pod up to 12" long. It is long-stalked and constricted between the bean-like brown seeds. The tree is very attractive and is easily propagated from seeds or seedlings.

The specimen here was photographed in Flower on Marathon Key in May. The seed pods picture was taken on Johnston Key in July.

SCIENTIFIC SYNONYM: *C. jamaicensis* Jacq.

Capparis flexuosa (L.) L. CAPPARACEAE
Common Names: **Limber Caper, Bay-leaved Caper Tree.**

This shrub or small tree is native to Florida, the Keys, the West Indies, the Bahamas and Tropical America. Its stems are unarmed and smooth. Its trunk frequently grows in a reclining posture to a height of 20' or more.

The leaves are light-green above and yellow-green below, alternate, blunt-pointed or notched at the apex, with unbroken margins, linear to broadly obovate in shape, glabrous and 2 to 6" long. The flowers have 4 white or pale pink petals with many long stamens. They are fragrant, showy and occur in axillary or terminal panicles. The fruits are greenish-yellow pods 6 to 10" long which split open when ripe to show a red interior and white seeds.

The above plant was photographed in a hammock on Islamorada in June.

Carica papaya L.   CARICACEAE
Common Names: **Papaya, Melon-tree.**

This tree-like herb is native to Central and Tropical America and the West Indies. It has become naturalized and grows wild throughout the Keys. The yellow, melon-shaped fruit is much smaller in the wild than the cultivated varieties. It is usually about 3" long and up to 2" in diameter. It is edible raw or cooked and is very sweet.

The leaves are very large and coarse with 7 to 9 deeply cut lobes growing on long stems. They contain papain and can be used as a meat tenderizer. Young leaves may be cooked as greens. The female flowers are greenish-yellow and are usually singular in the leaf axils. They are funnel-shaped with flaring petals. They are about 1" long and 1" wide. The male flowers appear in long-stem clusters on different trees. They may be bi-sexual and produce fruit. They bloom continuously and the fruit ripens throughout the year.

The specimens shown here were found on Sugarloaf and Summerland Keys.

Casasia clusiifolia (Jacq.) Urb.   RUBIACEAE
Common Name: **Seven-year Apple.**

The shrub or small tree reaching a height of 20' or more is native to South Florida and the Keys. It has pale bark and evergreen, leathery, 2 to 6" long leaves clustered at branch tips. They are glossy-green above and lighter below with curved under margins.

The flowers are white and tubular with star-like flared petals up to 1" long and 1½" wide, sometimes tipped with pink. They are fragrant and occur in axillary clusters throughout the year. The fruit is egg-shaped, 2 to 3" long and 2" wide. It is green and hard, turning yellow and finally shrinking and turning black like a prune when ripe. The dark-brown pulp is edible but not very tasty. It ripens in about one year — not 7 — as the common name implies.

It was photographed on Sugarloaf Key in May.

SCIENTIFIC SYNONYM: *Genipa clusiifolia* (Jacq.) Griseb.

Cassia chapmanii Isely   FABACEAE
Common Name: **Bahama Senna.**

This sprawling or upright shrub which reaches 7' or more in height is native to Tropical America, the West Indies, the Bahamas and South Florida. It is quite common on some of the Keys where it grows on the edges of hammocks among the mangroves.

It has light-green compound pinnate leaves with 5 or fewer pairs of oval leaflets from ½ to 2¼" long. the flowers are of the typical butterfly shape, about 1" across, in upper-leaf axillary clusters of 4 to 8 flowers and golden yellow in color. The flat, brown fruit pods are from 1 to 3" long and contain several shiny, brown seeds.

This shrub has been observed in flower in the Keys from October to July. It is very salt tolerant and may be easily propagated from seed.

The photographs were taken on Sugarloaf and Cudjoe Keys in June.

SCIENTIFIC SYNONYM: *C. bahamensis* Mill.

Cassia keyensis (Pennell) Macbride   FABACEAE
Common Names: **Big Pine Partridge Pea, Key Cassia.**

This shrub is endemic to the Florida Keys and is known to exist on Big Pine Key, No Name Key and Cudjoe Key. It is a freely branching, sprawling, prostrate and woody shrub. The pinnately compound leaves are alternate and 1″ or more long with up to 8 pairs of opposite, green, oblong or obovate leaflets ¼ to ½″ long.

The flowers are about 1″ across, have 5 golden-yellow, flaring petals with small, orange markings near the base and grow out of the leaf axils. They are usually singular. The fruit is a slender, pea-shaped pod up to 2″ long and contains 8 to 10 hard, small seeds.

The plant is usually found on the edges of hammocks and pinelands. The specimen shown here was photographed on Big Pine Key in July.

SCIENTIFIC SYNONYM: *Chamaecrista keyensis* Pennell.
DESIGNATED STATUS: Endangered.

Casuarina equisetifolia L. CASUARINACEAE
ex J.R. & G. Forst.
Common Names: **Australian Pine, Beefwood, Horsetail Tree, She-oak.**

This tree is a native of Australia but it has become naturalized in Tropical America, the West Indies, South Florida and especially in the Florida Keys. It spreads rapidly with a wide-ranging root system and easily sprouting seeds. It grows very fast making it very difficult for many of the slow growing native trees to compete for survival. In fact, it is classified as a pestiferous tree.

The local name "Australian Pine" is a misnomer. It is not a pine. The so-called needles are not needles but are jointed branches. There are whorls of 6 to 8 very small leaves (or scales) surrounding the nodes of the branches. The staminate spikes are up to 2" long. The pistillate spikes are woody, cone-like and about ½" in diameter. The tree may reach a height of 100'.

The flowering picture was taken on Saddlebunch Key in April.

SCIENTIFIC SYNONYM: *C. littorea* L.

Casuarina glauca Sieber ex K. Spreng.   CASUARINACEAE
Common Names: **Brazilian Oak, Scalybark Beefwood, Swamp She-oak.**

This tree is a native of Australia and has become naturalized in South Florida and the Florida Keys. It is similar to the *C. equisetifolia* except it has 10 to 16 of the small leaves (scales) in the whorls surrounding the nodes of the branches instead of 6 to 8. It does not normally grow as tall. The branches are much more dense and glossy-green with almost a bluish cast, giving it a soft brushy appearance. It spreads readily from root sprouts and forms dense thickets.

These pictures were taken on Vaca Key in February.

SCIENTIFIC SYNONYM: *C. lepidophloia* misappl.

Catesbaea parviflora Sw.   RUBIACEAE

Common Names: **Small-flowered Lily-thorn, Dune Lily-thorn.**

This small, spiny shrub is native to the Antilles and is found in the Bahamas, the West Indies and the Florida Keys. It is usually less than 2' tall but is reported to occasionally reach a height of 6' or more. The stems are stiff with very few branches and are covered with yellow-green, shiny, obovate, entire-marginal, opposite leaves about ½" long. In the leaf axils are two sharp green spines with dark tips that extend beyond the leaves.

The flowers are individual on short stems growing out of the leaf axils. They are white and funnel-shaped with 4 flaring petals. The fruit is a small, white berry with few seeds.

It was photographed on Big Pine Key in October. It is also found on Bahia Honda Key in the State Park.

DESIGNATED STATUS: Endangered.

Cereus gracilis var. *simpsonii*   CACTACEAE
(Small) L. Benson
Common Name: **Prickly Apple.**

This cactus plant is found in the Ten Thousand Islands area of South Florida and the Florida Keys. Its slender, 9 or 10 ridged stems may be simple or branched, erect or reclining, and reach a height up to 10'. It grows in either small groups or large clustered colonies. Each areole contains 7 to 9 spines, the longest of which is about 1".

The flowers are long and funnel-shaped with a swollen base, a long, scaled floral tube and white petals. The young buds are frequently clothed with white hairs, a distinguishing characteristic of this variety. The fruit is a depressed globular berry 1½ to 2½" in diameter. As it ripens it changes from yellow to a dull-red color at maturity. The seeds are about ⅛" long.

This plant was photographed on Big Pine Key in the Cactus Hammock in August.

SCIENTIFIC SYNONYM: *Harrisia simpsonii* Small.
DESIGNATED STATUS: Endangered.

Cereus pentagonus (L.) Haw.   CACTACEAE
Common Names: **Dildo Cactus, Barbed-wire Cactus.**

This night-blooming cactus is native to South Florida, Tropical America and the Florida Keys. The branches are usually 3-angled and reclining or, frequently, creeping over other plants up to 3 or 4' high. The branches or stems are jointed and can reach a total length of 20' or more. They are dark green in color and have tufts of 4 to 7 slender stiff spines. The middle ones are up to 1½" long and are rose colored on new growth.

The flowers are 3" or more in diameter, on 3 to 6" long tubes and have white petals with yellow centers. They open after dark and fade after sunlight strikes them in the morning. The fruit is bright red, pear or oblong in shape, 1½ to 2½" long, juicy and contains many small black seeds. It can be eaten raw.

The plant was photographed in flower on Big Pine Key in September and October.

SCIENTIFIC SYNONYM: *Acanthocereus floridanus* Small.
DESIGNATED STATUS: Threatened.

Cereus robinii (Lem.) L. Benson
var. *deeringii* (Small) L. Benson

CACTACEAE

Common Names: **Tree Cactus, Column Cactus.**

This upright columnar cactus with 9 or 10 deeply cut ribs grows to a height of 15 to 20'. The fleshy succulent trunks and branches are 4" thick and dull grayish green. Although it has no glochids, it does have 9 to 25 long, yellow spines in areoles along the ridges of the ribs. It is a night blooming cactus, the rare flowers opening after dark and wilting in the morning sunlight. They are bell-shaped, 1½ to 2" in diameter and on 3" fat bluish stems. The outer petals are purplish shading to green and finally to white in the center. There is a faint garlic odor to the flowers. The fruit is an oblong drupe about 1¼" in diameter and 2" long containing fleshy pulp and many small, hard seeds.

This plant is considered rare and endangered. Although native to the Keys, the only known plants still in existance are in "Cactus Hammock" on Big Pine Key, which is now a part of the Great White Heron National Wildlife Refuge. The flowers were photographed there in September.

DESIGNATED STATUS: Endangered.

Chiococca alba (L.) A.Hitchc.   RUBIACEAE
Common Names: **Snowberry, Snakeroot, Tears-of-St. Peter.**

This plant is native to South Florida, the West Indies and the Florida Keys. It is an upright shrub to about 10'. It has a sprawling habit with several trunks or it climbs vine-like on other trees. It has yellowish-gray bark. It is usually found in the hardwood hammocks.

The leaves are bright-green, opposite, sharp pointed, evergreen, leathery, elliptical or oval in shape, with entire margins and from 2 to 4" long. The flowers are creamy-white or yellow, bell-shaped, ¼ to ½" long and fragrant. They have a 5-lobed corolla and appear in axillary racemes. The fruits are white globular drupes, ¼ to ½" in diameter and have a mealy pulp surrounding the seeds.

This plant is very similar to the ***Chiococca parvifolio***, which is usually found in the pinelands. The distinguishing characteristics are pointed out in that plant's description.

This plant was photographed in flower on Plantation Key in June and on Cudjoe Key in August.

Chiococca parvifolia Wullschl. ex Griseb. RUBIACEAE
Common Name: **Snowberry.**

This plant is very similar to the ***Chiococca alba***. It is native to South Florida, the West Indies and the Florida Keys. It is a low trailing shrub with many vine-like branches 3 to 4' long.

The larger leaves are 1 to 2" long whereas the *C. alba* leaves are 2 to 4" long. The flowers are white or purplish-white. The *C. alba* has creamy-white or yellow flowers. The fruits of the *C. alba* are larger. In other characteristics the plants are about the same.

This plant was photographed on Sugarloaf Key in May and June.

SCIENTIFIC SYNONYM: *C. pinetorum* Britton.

Chrysobalanus icaco L. CHRYSOBALANACEAE
Common Names: **Coco-plum, Icaco Coco-plum.**

This shrub or small tree usually grows in thickets, has an upright or reclining trunk and may reach a height of 20'. It is native to South Florida, the Keys and the West Indies. Its sweet juicy fruit is eaten raw or made into jelly or preserves. It was a very important food for the Seminole Indians.

The dark-green leaves are oval or nearly round and slightly notched at the apex. They are alternate, leathery, glossy and 1 to 3½" long. New growth is reddish. The small white 5-petaled flowers are borne in short axillary clusters. The fruit is globular, 1-seeded, 1 to 1¾" in diameter and dark purple or white when ripe. The nut-like seed has an edible kernel.

The plants shown here were photographed on Big Pine Key in flower in October.

SCIENTIFIC SYNONYMS: *C. interior* Small, *C. pellocarpus* Meyer.

Chrysophyllum oliviforme L.　　　　　　SAPOTACEAE
Common Name: **Satin Leaf.**

This evergreen shrub or small tree, growing to 25 or 30', is native to South Florida, the Keys and the West Indies. The ripe fruit is eaten raw or made into jelly.

It has an upright trunk and red-brown bark. The leaves are oval in shape, dark green on the top side, shiny, 2 to 5" long, short pointed at the apex and covered with a golden-brown, satiny fuzz on the underneath side.

The flowers are tiny, light green, 5-lobed and borne in clusters in the leaf axils. The fruit is a dark purple, oval berry and ¾" long with flesh that is sweetish, juicy, lavender-purple in color and somewhat gummy.

The illustrative specimens are on Plantation Key and Islamorada and were photographed in March.

DESIGNATED STATUS: Threatened.

Cienfuegosia yucataniensis Millsp. MALVACEAE
Common Name: **None known.**

This shrub or subshrub is found in coastal hammocks in the Florida Keys, the West Indies and Northern South America. It has erect branching woody stems that may reach a height of 3' or more.

The upper leaves are lance-ovate in shape, alternate, have entire or lobed margins and are 1 to 1½" long. The lower leaves are generally longer, to 2½", and are broadly ovate or elliptic in shape. The flowers appear individually, have 5 sepals, 5 bright yellow petals and are about 1" across. The fruits are 3- to 5-valved capsules.

This plant was photographed on Windley Key in August.

Citharexylum fruticosum L.   VERBENACEAE
Common Names: **Fiddlewood, Florida Fiddlewood.**

This shrub or small tree, 10 to 25' tall, is native to South Florida, the Keys and the West Indies. It is evergreen with opposite, leathery, oblong-obovate, pointed leaves 3 to 6" long and 1 to 1½" wide. They are bright green and shiny above and dull, pale green below. The margins are entire, thickened and at times turned backward.

The fragrant flowers are white, tubular, 5-lobed and appear in axillary raceme clusters about 2 to 4" long. The fruit is globular in shape, ¼ to ½" in diameter and red-brown in color. It contains 2 stones and 4 seeds.

The tree has smooth, light-brown bark. Its wood is heavy, very hard, close grained with bright-red heartwood and much lighter colored sap wood.

The above specimens were photographed in flower on Big Pine and Sugarloaf Keys in June and July.

SCIENTIFIC SYNONYM: *C. villosum* Jacq.

Citrus aurantifolio (Christm.) Swingle   RUTACEAE
Common Name: **Key Lime.**

This tree is native to Southern Asia and has become naturalized in the Florida Keys. In fact, the Keys have become known for the Key lime pies made from its fruit. It is a shrubby tree with thorny branches growing to a height of up to 15'.

The leaves are elliptic-ovate in shape, shiny, dark-green, aromatic, evergreen, with crenate margins and from 2 to 3" long. The petioles are frequently narrowly winged. The flowers have waxy-like petals and many stamens. They appear in axillary clusters throughout the year.

The fruit is globular and from 1 to 2" in diameter. It is yellow when ripe, although many people prefer to use it in the green stage in drinks and pies.

The plant was photographed on Sugarloaf Key in December.

SCIENTIFIC SYNONYM: *C. lima* Lunan.

Clusia rosea Jacq.   CLUSIACEAE

Common Names: **Autograph Tree, Pitch Apple, Balsam Apple, Signature Tree, Monkey Apple, Scotch Attorney, Copey.**

Native to the Bahamas, the West Indies, Tropical America and the Florida Keys, this plant has wide-spreading horizontal branches. A seed may sprout on another tree and grow as an epiphyte, sending down aerial roots to the ground and eventually strangling the host tree like the Strangler Fig. It is considered very rare in the wild state.

The leaves are thick, persistent, opposite, dark glossy-green, obovate, rounded and notched at the apex, up to 5" wide and 7" long. If its surface is scratched by pen or nail, the leaf will develop a scar and retain it even after it is removed from the tree—thus the common name "Autograph Tree."

The flowers are beautiful, about 3" across, with white petals tinted pink around a yellow center and a definite porcelain luster. The 3" seed pod is equally beautiful when it opens, claw-like with 8 to 12 sections exposing hard, oval, black seeds in red pulp.

The plant pictured here was found growing wild in a hammock on Cudjoe Key in September. It is 25' tall.

DESIGNATED STATUS: Endangered.

Coccoloba diversifolia Jacq.   POLYGONACEAE
Common Names: **Pigeon Plum, Dove Plum.**

A native to the Keys, South Florida and the West Indies, this plant has a straight trunk and a rounded compact leaf. The above specimen on Sugarloaf Key is 32' high. It has light gray bark tinged with brown. As the botannical name *(diversifolia)* implies, the leaves are very diversified in shape and size varying from 2 to 5" long. They are leathery, alternate, bright green above and paler below, with clasping petioles like the Seagrape. The flowers are inconspicuous, without petals, a creamy white, cup-shaped calyx on 2 to 3" long spikes. The fruits are dark purple, thin fleshed, round or pear-shaped and have a ¼ to ½" single hard seed. The fruit is juicy, acid and somewhat astringent. It is edible raw or made into jelly or wine.

The flowers were photographed in April and May on Sugarloaf Key.

SCIENTIFIC SYNONYMS: *C. floridana* Meissner, *C. laurifolia* Jacq.

Coccoloba uvifera (L.) L.   POLYGONACEAE
Common Names: **Sea-grape, Shore-grape, Seaside-grape.**

Native to the Florida Keys, South Florida and the West Indies, it grows as a sprawling shrub or a small tree to 20'. The leaves are evergreen, almost round and 4 to 8" in diameter. Young leaves are bronze, turning green and finally bright red before falling. The flowers are inconspicuous, yellowish-green to white, on slender racemes, 6 to 12" long and fragrant. The fruit, in 5" or more grape-like clusters, are pear shaped about ¾" long. They turn red to purple as they ripen one grape at a time. The ripe fruit is edible raw, can be used to flavor meat and makes excellent jelly.

The ripe fruit falling on the ground is very attractive to bees.

The specimen pictured here was photographed on Sugarloaf Key in April.

Coccothrinax argentata (Jacq.) L.H. Bailey   ARECACEAE
Common Names: **Silver Palm, Florida Silverpalm, Biscayne Palm, Seamberry Palm.**

This palm tree is native to the Florida Keys, South Florida and the West Indies. It has a slender, smooth, brownish-gray trunk and may reach a height of 20'. The deeply-divided leaves are fan-shaped, about 24" across, glossy-green above and silvery gray beneath. The unarmed leaf stalks may be 3' long and flexible. The flowers, borne in 2' clusters, are very small, ivory-white and fragrant. The fruit is round, about ½" in diameter, with a single deeply grooved seed, red turning purple or black when ripe. It is edible but not very tasty.

The leaves are often woven into hats, baskets and mats. The terminal bud may be cooked like that of the Cabbage Palm.

The fruited specimen is on Cudjoe Key and was photographed in September.

SCIENTIFIC SYNONYMS: *C. garberi* Sarg., *C. jucunda* Sarg.
DESIGNATED STATUS: Endangered.

Cocos nucifera L.   ARECACEAE
Common Name: **Coconut Palm.**

Most authorities agree the coconut palm originally came from Asia or the islands of the Indian Ocean. Coconut palms were reported growing in Key West in 1824 and on Boca Chica and Sugarloaf Keys in 1877.

It has a slender trunk, bulbous at the base, often leaning or bowed, topped by a rosette of pinnate green leaves up to 20' long. The flowers are small and light yellow in plume-like clusters breaking out of a long, pointed sheaf in the leaf axils. Healthy trees bloom continuously all year. The fruits grow to a 3-sided thick fibrous husk 12 to 15" long. It contains an oval-shaped nut with a hard shell and a ½" lining of white meat and up to a pint of sweetish liquid. The ripe meat is eaten in its natural state or mixed with candy and pastries. It is also used in making cooking oil, skin oil, soap and many other products.

Varieties—"Golden," "Yellow," and "Green"—shown above were photographed on Sugarloaf Key in June.

Colubrina arborescens (Mill.) Sarg. RHAMNACEAE
Common Names: **Wild Coffee, Coffee Colubrina, Snakebark.**

This species is native to South Florida, the Bahamas, Central America, the West Indies and the Florida Keys. It is a small tree or shrub which may reach a height of 20' or more. The young stems are covered with rusty hairs. It has elliptic or ovate-lanceolate leaves, 2 to 4" long with entire margins, dark green and lustrous above and some rusty pubescence below with scattered black glands.

The flowers are small in axillary clusters and yellow in color. The fruits are subglobose, about ¼" in diameter and purplish to black in color when ripe, which at times rupture explosively to release three black seeds.

It was photographed in flower on West Summerland Key in February.

SCIENTIFIC SYNONYM: *C. colubrina* (Jacq.) Millsp.

Colubrina asiatica (L.) Brongn.　　　　　　RHAMNACEAE
Common Names: **Colubrina, Latherleaf.**

This very attractive bushy shrub is a native of the Old World beaches. It has become naturalized in South Florida, the West Indies and the Florida Keys. It has slender stems that may be spreading, trailing or erect to 10' high.

The leaves are dark green above, lighter below, glossy, alternate, ovate or elliptic-ovate, 3-nerved at the base, with long soft points at the apex, serrate margins and up to 4" long. The flowers are bisexual, small, 5-petaled, flaring, greenish and in axillary clusters. The fruit is a spherical drupe about ¼" in diameter. The seeds sprout readily and the shrub spreads rapidly. It forms dense thickets and can crowd out other plants.

The plant was photographed on Geiger and Sugarloaf Keys. The flower was photographed in May.

Colubrina elliptica (Sw.) Briz. & Stern   RHAMNACEAE
Common Names: **Soldierwood, Naked-wood.**

This tropical hardwood tree is native to the Bahamas, the West Indies, Mexico, Tropical America and the Florida Keys. Although usually of shrub size, it can reach a height of 30 to 50'. The orange-brown bark is marked by deep serpentine furrows. The ovate to lanceolate or elliptical leaves are soft, weak, from 2 to 4" long, 1½ to 2" wide, rounded or gradually tapered at the base and abruptly tapered to a blunt point. There are small marginal glands at the base of the leaf blades.

The flowers are yellow or greenish-yellow, 5-petaled, about ¼" across and borne in axillary clusters. The fruit is an orange-red clustered capsule about ¼" in diameter and it contains 3 oblong, dark seeds. The dry fruits burst open with a recognizable "pop" when heated by the sun. This gives them their common name "Soldierwood" because it sounds like musket-fire. The pop is actually nature's way of scattering the seeds.

The plant was photographed on Long Key in July.

SCIENTIFIC SYNONYM: *C. reclinata* (L.Her.) Brongn.

Conocarpus erectus L. COMBRETACEAE
Common Names: **Buttonwood, Button-mangrove.**

Probably the most prominent shoreline tree in South Florida, the West Indies and the Florida Keys is the "Buttonwood." It grows as a shrub or a wide-crowned tree to a height of 30' or more. The dense, hard wood was used extensively during the "Sailing Ship" days in the manufacture of charcoal. It is still a favorite for smoking fish and other meats.

The leaves are evergreen, smooth, alternate, ovate, obovate or elliptical in shape with entire margins and 1 to 4" long. The flowers are very small, greenish, without petals and occur in dense, cone-like heads. The fruits are purplish-green cones.

This plant is an excellent choice for water front gardens because of its high tolerance for salt, wind, wet or dry seasons, full sun or partial shade.

This tree was photographed on Sugarloaf Key in January.

Conocarpus erectus L. var. *sericeus*　　COMBRETACEAE
Forst. ex DC.
Common Names: **Silver Buttonwood, Silver-leaved Buttonwood.**

This variety is very similar to the ***Conocarpus erectus*** except it has silver-gray pubescent leaves. It is found generally in the same areas.

The silver variety is widely used as a landscape plant by homeowners and developers. It not only is very attractive, but being tolerant of climatic conditions in the Keys, it requires little attention after it has become established.

The above plants were photographed on Sugarloaf Key in January.

Cordia globosa (Jacq.) HBK.   BORAGINACEAE
Common Name: **Bloodberry.**

This shrub is native to South Florida, the West Indies and the Florida Keys. It is well-branched and may reach a height of 8 to 10'. It is usually found growing in the hammocks.

The leaves are small—¾ to 1½" long. They are elliptic to lanceovate in shape with toothed margins. They are persistent, petioled and covered with hairy fuzz giving them a grayish green color.

The white flowers appear in scorpioid terminal clusters. The corolla is funnelform in shape and about ¼" long. The fruit is a red drupe partly enclosed in the calyx.

The plant was photographed in the vicinity of Fort Taylor in Key West in August with fruit and on Windley Key in flower during the same month.

SCIENTIFIC SYNONYM: *Varronia globosa* Jacq.

Cordia sebestena L.   BORAGINACEAE
Common Names: **Geiger Tree, Scarlet Cordia, Geranium Tree, Vomitel.**

Native to the Florida Keys, South Florida and the West Indies, this tree has dark brown bark, a dense, rounded crown and grows about 20' tall. Its leaves are rough, evergreen, broad oval with pointed tips, 5 to 10" long and 3 to 5" wide, dark dull-green above and lighter below, in terminal clusters. The flowers are bright red or orange, 1 to 1½" wide, flared-trumpet shaped, in flat terminal clusters. It blooms practically all year. The fruit, white when ripe, is pear shaped, fleshy but fibrous, fragrant and sweetish, 1 to 1¼" long, containing 1 or 2 seeds. It can be eaten raw or cooked but it is not very tasty.

A common local story claims John James Audubon named the tree for Captain Geiger, a successful Key West wrecker with whom he lived during a stay in the Keys in what is now known as the Audubon House.

Photographed in May on Bahia Honda Key in flower.

SCIENTIFIC SYNONYM: *Sebestena sebestena* L.
DESIGNATED STATUS: Threatened.

Crescentia cujete L. BIGNONIACEAE
Common Name: **Calabash Tree.**

Although this tree is considered by some to be native to the Florida Keys as well as the West Indies and Tropical America, research for this book found no specimens growing wild. Those photographed were in private gardens in Key West and Sugarloaf Key.

It has spreading branches that result in a wide crown and reaches a height of 25 to 30′. The leaves are up to 6″ long, spatulate to oblanceolate in shape, evergreen and clustered with 3 to 5 leaves in a bundle.

The flowers are 5-petaled, bell shaped, greenish-yellow with purple streaks, about 2½″ wide and grow directly along the trunk and branches. They open after dark and drop off in the early morning. They are a favorite of honey bees that gather in large numbers on the fallen flowers. The fruits are subglobose to ellipsoid, 4 to 8″ in diameter, hard-shelled and green, turning brown as they ripen.

The flower pictures were taken in March.

Crossopetalum ilicifolium (Poir.) Kuntze   CELASTRACEAE
Common Names: **Ground-holly, Quail-berry, Christmas-berry.**

This evergreen shrub is found in the West Indies, Central America, South Florida and the Florida Keys. It differs from the ***Crossopetalum rhacoma*** generally in three ways:

(1) It grows as a prostrate sprawling shrub instead of an erect tree. (2) The leaves are definitely sharply toothed or spiny, similar to the American Holly *(Ilex opaca)*. (3) The leaves are much smaller—usually ⅘" long or less.

The fruit is an orange-red spherical drupe about ¼" in diameter. The plant is usually found growing along the edges of hammocks.

The plants were photographed on No Name and Sugarloaf Keys in July.

SCIENTIFIC SYNONYM: *Rhacoma ilicifolia* L.

Crossopetalum rhacoma Crantz — CELASTRACEAE
Common Names: **Rhacoma, Florida Crossopetalum.**

The Rhacoma is an evergreen shrub or small tree, native to South Florida, the West Indies and the Florida Keys. It may reach a height of 15 to 20', but is usually much smaller. It normally grows erect and has smooth, light-brown bark.

Its leaves are opposite, alternate or whorled. They are light-green above and paler below with inconspicuous veins, oval-obovate in shape, ½ to 1½" long and ¼ to ½" wide. The margins are shallowly crenate or entire and the apex is either rounded or slightly notched.

The flowers are very small, long stalked, in thin clusters growing out of the leaf axils. The fruits are red or reddish purple drupes about ¼" in diameter containing a bony stone and very small seeds.

It was photographed on Sugarloaf Key in flower in January.

SCIENTIFIC SYNONYM: *Rhacoma crossopetalum* L.

Croton linearis Jacq.   EUPHORBIACEAE
Common Names: **Pineland Croton, Wild Croton, Granny-bush.**

This shrub is native to the pinelands and coastal areas of South Florida, the West Indies and the Florida Keys. It usually grows in clumps and may reach a height of 6' or more.

The leaves are narrowly linear from 1½ to 3" long. They are dark green and smooth above and yellowish pubescent below. Reportedly they are sometimes used for making tea. The author has never tried it. The staminate flowers appear on racemes up to 4" long. They have approximately 15 stamens and are separate from the pistillate flowers but both occur on the same plant. The fruits are subglobose, yellowish, pubescent capsules containing small seeds.

The specimen shown here was photographed on Big Pine Key in May.

Cupania glabra Sw.   SAPINDACEAE
Common Names: **Cupania, Florida Cupania.**

This small to medium size tree is native to the Florida Keys, Tropical America, Jamaica and Cuba. *The Atlas of United States Trees* published by the United States Department of Agriculture Forest Service in 1978 classifies this plant as very rare and in the United States found only in the Lower Florida Keys.

It is evergreen and the leaves are alternate and pinnately compound with 6 to 12 oblong round-tipped leaflets 2½ to 8″ long. The leaflets have low rounded teeth in the margins. The flowers are very small in axillary panicles that are about the same length as the leaves. They may have 5 white petals or none. The fruits are 3-lobed capsules about ¾″ long containing 3 round black seeds.

The specimen shown here were photographed on Big Pine Key in fruit in May and on Summerland Key in flower in November.

DESIGNATED STATUS: Endangered.

Dalbergia brownei Chapm.   FABACEAE
Common Name: **Coin Vine.**

This sprawling, woody shrub is found in Central and Tropical America, the West Indies and the Florida Keys. It may reach a height of 6' or spread over a large area 1 to 2' above the ground or climb above other plants forming dense thickets several yards wide. The leaves are compound with leaflets arranged alternately along the stems. The leafters are glabrous, ovate to elliptic in shape, dark green, leathery and from 1 to 4" long. New leaflets are soft and yellow-green.

The small flowers have a white or pinkish corolla and occur in the axils of the branches. The fruits are flat pods with a circular outline about 1" in diameter containing a single seed which readily floats on the tide aiding in the distribution of the plant along shorelines. The fruits become copper colored at maturity— resulting in the common name "Coin Vine."

This plant was photographed on Upper Key Largo in flower in March.

SCIENTIFIC SYNONYMS: *D. amerimnon* Benth., *Amerimnon brownei* Jacq.

Dodonaea viscosa (L.) Jacq.   SAPINDACEAE
Common Names: **Florida Hop Bush, Hop Bush, Varnish-leaf.**

A native of the Florida Keys, South Florida and the West Indies, this plant grows either as a shrub or a small tree to 15'. The leaves are either narrow-elliptic or spatulate in shape and from 1 to 6" long. The blunt apex is rounded or faintly notched, dark green, shiny with a resin coat and evergreen with curled under edges.

The flowers are small, without petals, greenish yellow and in terminal clusters. The 3-celled seed capsule with 3 to 4 wings contains 1 or 2 seeds in each cell. The capsule is yellowish, turning red and finally brown when dry.

The resin coating on the leaves protects them during droughts, helping them to remain fresh when other plants are wilting.

This specimen was photographed on Sugarloaf Key in December with fruit and on Key Largo in flower in October.

SCIENTIFIC SYNONYMS: ***D. jamaicensis*** DC., ***D. microcarya*** Small.

Drypetes diversifolia Krug & Urb.   EUPHORBIACEAE
Common Names: **Milkbark, Whitewood.**

This shrub or small tree, growing to 30' or more, is native to South Florida, the Keys, the Bahamas and the West Indies. The bark is milk-white and is frequently marked by colorful lichens. The leaves are dark green and, as the species name *(diversifolia)* implies, occur in different sizes and shapes. They are usually oblong or oval, 3 to 5" long, 1 to 2" wide, rounded at the apex and sometimes minutely notched. They are leathery, alternate and persistent. The leaves of young seedlings have spines in the margins. The flowers are tiny in dense clusters in the leaf axils. They have a 5-lobed calyx but no petals.

The fruits are ½ to 1" ivory-white ovoid drupes with a dry mealy pulp covering a single stone and containing 1 seed.

Since the hard, heavy wood seemed to be immune from attacks by the shipworm, the natives frequently used it for wharf pilings.

The specimens included here were photographed on Sugarloaf and Long Keys. The flower picture was taken in April.

SCIENTIFIC SYNONYM: *D. keyensis* Krug & Urb.

Drypetes lateriflora (Sw.) Krug & Urb.   EUPHORBIACEAE
Common Name: **Guiana Plum.**

This small tree or shrub is considered rare in the Florida Keys and North along the coast to Broward County. It also appears in the Bahamas, the West Indies and Central America. Its shiny, alternate, leathery leaves with entire margins and raised veins are from 3 to 4" long. They are abruptly pointed, lanceolate to ovate in shape and often oblique at the leaf bases.

The plants are dioecious. The small flowers in axillary clusters are about ⅛" wide and have 4 or 5 sepals and 3 to 12 stamens. There are fewer flowers in the pistolate clusters. The fruit is a red, downy drupe, nearly globular in shape and ¼" or less long.

The plant shown here was photographed in a hardwood hammock on Long Key in February.

Duranta repens L.   VERBENACEAE

Common Names: **Golden Dewdrop, Pigeon-berry, Sky-flower.**

Some authors have recorded this shrub or small tree as being native to the Florida Keys, South Florida, the Bahamas, the West Indies and Tropical America. The research for this book found it growing only in or near established gardens in the Keys. It is planted frequently as an ornamental. It has slender stems, is occasionally spiny and may reach a height of 15 to 18'.

    The leaves are 1 to 3" long, evergreen, opposite, ovate to elliptic in shape, of medium texture, with serrated or entire edges, obtuse or acute apexes, and growing on short petioles. The flowers are small, about ½" across, with 5-lobed cylindrical corollas varying from white to blue to purple. They appear on loose axillary racemes from 5 to 6" long. The fruits are orange-yellow globose drupes from ¼ to ½" in diameter containing 4 nutlets. They are supposedly poisonous to humans but a choice food for birds.

    The plants were photographed in flower from June to September on Sugarloaf Key and Stock Island.

Erithalis fruticosa L.   RUBIACEAE
Common Names: **Black Torch, Erithalis.**

A native to South Florida and the Keys, this densely leaved shrub is an excellent fill-in between ground covers and the taller hammock hardwoods. It grows to about 8' in height and has dark-brown bark striped with light brown. The leaves are thin, flat, medium green above and lighter below. They are smooth with obscure veins, oval in shape, blunt-pointed, opposite and 1 to 2" long.

The flowers are small with star-shaped, white, 5-petaled corollas on panicles out of the leaf axils. The fruits are nearly globular shaped berries about ¼" in diameter. They are green turning to dark purple or black when ripe.

The plants shown here were photographed on Big Torch Key in flower in October.

Ernodea littoralis Sw.   RUBIACEAE

Common Names: **Golden-creeper, Beach-creeper.**

This is a vine-like shrub with long curving prostrate or spreading branches. It is native to South Florida, the Keys and the West Indies. Its narrowly-elliptic leaves have a yellow cast which explains the common name "Golden Creeper." The leaves are fleshy, leathery and 1" or less in length. They generally grow in clusters.

The flowers that grow out of the leaf axils are small, tubular with 4 to 6 lobes and are pink or yellowish with a long, white corolla. The fruits are small and usually round but sometimes oval. They are yellow and contain a single seed.

The plant is very tolerant of dry weather, grows in the sun and is usually found on coastal dunes and pinelands.

The plant shown here was photographed on Sugarloaf Key with flowers in November.

DESIGNATED STATUS: Threatened.

Erythrina herbacea L.   FABACEAE

Common Names: **Coral Bean, Cherokee Bean, Cardinal Spear, Eastern Coral Bean, Snakeweed.**

This native shrub or small tree grows to 15' high in the coastal hammock areas of Southeastern U.S. and the Florida Keys. The bark is furrowed, light gray, and the smaller branches frequently have spines. The leaves are deciduous but persistenl, alternate and compound with 3 medium-green arrowhead-shaped leaflets 1 to 3" long.

The bright red flowers in pyramid-shaped terminal racemes are closed, slender, tubular, about 2" long and frequently appear before the new leaves. The 3 to 6" long, drooping, stringbean-like seed pods, constricted between seeds, split open to reveal kidney-shaped hard, red seeds about ½" long. They are reputed to be poisonous and used in Mexico to poison rats and dogs. They are hard enough to be strung as beads.

This specimen was photographed in Watson Hammock on Big Pine Key in May.

SCIENTIFIC SYNONYM: *E. arborea* (Chapm.) Small.

Eugenia axillaris (Sw.) Willd.   MYRTACEAE
Common Names: **White Stopper, Wattle, Stopper, White Stopper Eugenia.**

This tall shrub or small tree, growing to about 20', is a native of South Florida and the Keys as well as the Bahamas and the West Indies. It is found in the hammocks and the sandy areas near the coast. Its evergreen leathery leaves are opposite, ovate, pointed, dark green above and paler, with fine black dots below. They are 1 to 3" long and aromatic. Some describe the odor as skunk-like. The new leaves are pink.

The small flowers have many stamens and 4 white petals. They are fragrant and occur in axillary racemes. The fruit is globular in shape, reddish-purple turning black, 1-seeded, sweet, juicy and less than ½" in diameter. The fruit is edible raw and the leaves supposedly have been used in the treatment of diarrhea, thus the common name "Stopper."

These photographs were made on Cudjoe and Sugarloaf Keys, the flowers in July and the fruit in January.

SCIENTIFIC SYNONYM: *E. monticola* DC.

Eugenia confusa DC. MYRTACEAE
Common Names: **Red-berry Stopper, Red-berry Eugenia.**

This tree is native to South Florida, the Bahamas, the West Indies and the Florida Keys. It is not common, however, at the present time in the Keys. It is easily recognized by its glossy, long, tapered-tipped, downward-pointed, stiff leaves with curled under margins. The leaves are opposite, elliptic or ovate in shape and from 1½ to 2" long.

It usually has a straight trunk with light-gray scaly bark and may reach a height of 18 to 20'. The flowers occur on the current growth in the axils of the leaves or basal scales. They have a corolla of white petals that is about ¼" wide. There are many stamens and approximately ¼" stems. The fruits are 1-seeded, subglose, brilliant red berries about ¼" in diameter.

The red-brown heartwood is close-grained, heavy and very hard. It is very desirable for fine cabinet work.

It was photographed on Key Largo in November.

DESIGNATED STATUS: Threatened.

Eugenia foetida Pers.   MYRTACEAE

Common Names: **Spanish Stopper, Box-leaf Stopper, Gurgeon Stopper.**

This shrub or small tree, which may reach a height of 15 to 20', is native to South Florida, the Keys, Central America and the West Indies. It has light brown bark scarred by old leaf bases. It is evergreen, the leaves being persistent and remaining on the tree until the second year. The leaves are aromatic, opposite, oval or elliptic in shape and from ½ to 2" in length. They have fine black dots on the underside.

The small white flowers appear in stalkless clusters along the branches. The globular fruit, which change from red to black as they ripen, are a favorite food for many native birds. They are about ¼" in diameter and contain thin flesh over 1 or 2 light brown seeds.

The specimen shown here was photographed on Sugarloaf Key in flower in August.

SCIENTIFIC SYNONYMS: *E. buxifolia* (Sw.) Willd., *E. myrtoides* Poir.

Eugenia rhombea (Berg) Krug & Urb. MYRTACEAE
Common Names: **Red Stopper, Spiceberry, Spiceberry Eugenia, Stopper.**

In the U.S.A. this shrub or small tree (to 15') is found only in the Florida Keys. The plant is also native to the West Indies and Central America. The bark is light gray and smooth. The wood is heavy, hard, close-grained; the heart and sapwood are both light brown in color, making it very suitable for furniture and wood carvings.

The evergreen leaves are dark green above, paler below, with black dots and faintly outlined, yellow margins, about 2½" long and 1 to 1¼" wide. They are ovate, tapering to a narrow rounded tip. New leaves are thin and light red. The flowers are white, up to ½" wide on clustered axillary stems. The nearly round or globular fruit is orange-red or, when mature, red turning black. It has thin dry flesh and brown seeds. The fruit is depressed at the apex.

The above pictures were taken in a Key West garden in an apparently preserved portion of a hardwood hammock. It was also found growing wild on Lignum Vitae Key. It was photographed in flower in March.

SCIENTIFIC SYNONYMS: *E. anthera* Small, *E. procera* Poir.
DESIGNATED STATUS: Endangered.

Exostema caribaeum (Jacq.) Roem. & Schult. RUBIACEAE
Common Names: **Princewood, Caribbean Princewood.**

This shrub or tree, which may reach a height of 20' or more, is native to the Bahamas, the West Indies, Tropical America and the Florida Keys. It has slender erect branches and heavy, hard, strong, close-grained, light-brown wood streaked with yellow. Supposedly the bark was once used to reduce fever; later quinine was used.

The leaves are medium green, leathery, shiny on the upper surface, 1 to 3" long, opposite, oblong-lanceolate with a sharply pointed tip and conspicuous mid-veins. The white or pinkish flowers turning orange are singular in the leaf axils and about 3" long. They have a flaring tube with 5 spreading lobes and 5 stamens. They are fragrant. The fruits, woody capsules about ½" long, turn black when dry and contain many winged brown seeds.

The flowering plant was photographed in a hammock on Islamorada in September.

Exothea paniculata (Juss.) Radlk.   SAPINDACEAE
Common Names: **Inkwood, Butter-bough, Ironwood.**

This shrub or tree is native to South Florida, the Keys and parts of Central America. It can reach a height of 40 to 50'. It has dark gray bark and shiny, green leaves clustered toward the end of the twigs. The leaves are pinnately compound with 2 or 4 oblong or oval leaflets, 4 to 5" long, rounded or minutely notched at the apex.

The flowers are fragrant, about ¼" across with 5 white petals in small panicles either axillary or terminal. The dark red fruit becomes purple when ripe, is globular in shape, ½" or more long, and contains 1 seed.

The wood is very hard, heavy, strong and close-grained. It is bright red-brown in color, the sapwood being a lighter shade. It is used for construction of small boats, tool handles and novelties.

The plant grows in the hammocks in limestone soil. The flower pictures were taken in Marathon on Vaca Key in March.

Ficus aurea Nutt.   MORACEAE

Common Names: **Strangler Fig, Wild Fig, Golden Fig, Florida Strangler Fig.**

Native to Southern Florida and the Keys, the Bahamas, and West Indies, this large tree reaches 50 to 60' high with a dense, broad, round top. It is epiphytal and at times a seed will sprout on another tree and send down aerial roots to the ground where they will join together forming a large trunk, eventually strangling the host tree—thus the name "Strangler Fig." It has smooth, gray or light-brown bark and milky sap.

The leaves are semi-deciduous usually remaining on the tree about 2 years. They are 2 to 5" long, dark yellow-green, shiny with yellow midribs, oblong or elliptical, usually narrowed at both ends and pointed at the apex. The flowers are small, in closed receptacles in the leaf axils, on very short stalks. The golden-yellow or bright-red, berrylike fruits about ½" in diameter contain small brown seeds.

The specimen shown here was photographed on Sugarloaf Key in June.

Ficus citrifolia Mill.   MORACEAE

Common Names: **Shortleaf Fig, Wild Banyan, Wild Fig.**

Native to the Florida Keys, the West Indies and the Bahamas, this tree grows in the hammocks throughout the Keys. It is a large tree to 50′ tall, with a broad top, light gray bark, some aerial roots and milky sap. Its leaves are dark green, lighter below, oval with a rounded base and an abruptly pointed tip. They are long-stalked, alternate, semi-deciduous and 2 to 5″ long.

The flowers are very small and enclosed in the open-ended fruit. The fruit, on fairly long stalks coming out of the leaf axils, are nearly round, slightly flattened at the end, ¾ to 1″ in diameter and change from yellow to dark red when ripe. They are sweet and edible raw.

The specimen here was photographed on Sugarloaf Key in April.

SCIENTIFIC SYNONYMS: ***F. brevifolia*** Nutt., ***F. laevigata*** Vahl, ***F. populnea*** Willd.

Forestiera segregata (Jacq.) Krug & Urb.　　　　　OLEACEAE
Common Names: **Florida Privet, Wild Olive, Florida Forestiera.**

The shrub or small tree, 10 to 20' high, is native to Florida, the Keys, the West Indies and Central America. It has pale gray, smooth bark and many branches forming a dense crown. The leaves are semi-deciduous in that they drop as the new leaves appear. They are opposite, dark green and shiny above, dull below, 1 to 3" long, elliptic to elliptic-spatulate, with blunt or pointed tips and entire margins.

The flowers are small, in clusters along the branches. They have no corolla but very attractive greenish yellow to white stamens. The fruit is a black, oval, thin-fleshed drupe from ¼ to ½" in length. It contains a single seed in a small olive-like stone.

This specimen was photographed in flower on Long Key in October.

Gossypium hirsutum L.   MALVACEAE

Common Names: **Wild Cotton, American Upland Cotton, Upland Cotton, Cotton, Common Cotton.**

This plant, native to that part of Tropical America that is frost-free, was once very common on the Keys. It is now on the "Endangered List." This change was due in part to the intentional efforts by Federal and State agencies to eradicate the plant because it was believed to be a potential contributor to the spread of the Pink Boll Worm to the commercial cotton plants.

This shrub may reach a height of 10' or more with spreading branches and coarsely pubescent stems. The leaves are opposite, 2 to 6" long and usually 3-lobed on long petioles. The pale yellow or creamy-white, hibiscus-like flowers fade to pink and have reddish spots on the base of each petal; they are 2 to 2½" wide. The fruit is a rough triangular-shaped capsule which splits open when ripe to expose seeds covered with white or brownish cotton.

It was photographed on Plantation Key in May in flower.

SCIENTIFIC SYNONYMS: *G. herbaceum* L., *G. punctatum* Schumach.
DESIGNATED STATUS: Endangered.

Guaiacum sanctum L. ZYGOPHYLLACEAE
Common Names: **Lignum Vitae, Holywood, Tree of Life, Roughbark Lignum Vitae.**

This is one of the most outstanding trees in the world. It has a very high resin content (about 30%) making it most valuable where non-contaminate lubrication is desired. Ship propeller-shaft bearings, food-handling machinery, pulleys, neck bearings, etc. make such a demand for the wood that the tree is on the "Endangered List." It is a native to South Florida, the Keys and the West Indies. A specimen in Key West is 21' high with a 22' spread and a trunk circumference of 26" at a point 4½' above ground level.

The leaves are evergreen and pinnate with 6 to 8 dark glossy leaflets about 1" long tipped with a small sharp point. The flowers occur at branch tips either singly or in clusters. They are blue, have 5 petals, are star-like and about ¾" wide. The 5-angled orange-yellow fruits split open to reveal black seeds with red arils.

The above specimens were photographed in flower in Marathon and Sugarloaf Key in August.

SCIENTIFIC SYNONYM: ***G. guatemalense*** Planchon ex Rydberg.
DESIGNATED STATUS: Endangered.

Guapira discolor (K. Spreng.) Little   NYTAGINACEAE
Common Names: **Blolly, Longleaf Blolly, Beefwood.**

The blolly is native to South Florida, the Keys and the West Indies. It is a large shrub or small tree reaching a height of 30 to 40'. It has light reddish-brown bark and many vertical branches. The thin, light-green leaves have obscure veins and thickened wavy margins. They vary in shape from oblong to obovate, ½ to 2½" long and ½ to 1" wide.

The small flowers have a tubular-shaped greenish-yellow or purplish calyx and no petals. They grow in terminal and axillary clusters. The red, juicy, 10-ribbed, oval fruit, ¼ to ½" in diameter, is reported to be edible. It encloses a light-brown cylindrical nutlet. The wood is heavy, soft, weak, coarse-grained and yellowish-brown.

The plant is usually found in hardwood hammocks or pinelands. The specimen photograph used here was made on Stock Island in July.

SCIENTIFIC SYNONYMS: *G. longifolia* (Heimerl) Little, *Torrubia longifolia* (Heimerl) Britton, *T. globosa* Small, *Pisonia discolor* Spreng.

Guettarda elliptica Sw.   RUBIACEAE
Common Names: **Velvet-seed, Elliptic-leaf Velvetseed, Everglades Velvetseed.**

This shrub or small tree may reach a height of 15 to 20'. It is native to South Florida, the Bahamas, the West Indies, Tropical America and the Florida Keys.

The leaves are usually less than 2" long and more or less glabrous or covered with very fine hairs on the underneath side. They are dull light-green in color. The flowers are white, pink or reddish, tubular in shape, about ¼" long with a 4- or 5-lobed corolla. They occur either solitary or in few-flowered clusters. The fruit is globular, flattened or slightly indented at the tip and red turning black as it ripens. It normally is about ⅖" in diameter.

This tree was photographed on No Name Key in July.

Guettarda scabra (L.) Vent.   RUBIACEAE
Common Names: **Rough Velvetseed, Roughleaf Velvetseed.**

Native to South Florida, the Keys and the West Indies, this shrub or small tree grows to about 15' tall. It is evergreen and has thick oval leaves 2 to 6" long. They are rough, leathery, dark green and sometimes have a short, pointed tip. They have fine hairs on both surfaces.

The flowers are white, tubular, 5-lobed, about 1" long and grow out of the leaf axils. The fruits are red to purple from ¼ to ½" in diameter, nearly globular, flattened and slightly indented on the end. They are sweet and mealy, with a very velvety skin and contain 4 to 6 seeds.

The illustrative specimen was photographed on Sugarloaf Key in flower in July.

Gyminda latifolia (Sw.) Urb.   CELASTRACEAE
Common Names: **False Boxwood, West Indies False-box.**

This large evergreen shrub or small tree is native to the West Indies, the Bahamas, Central America and the Florida Keys. In the hammocks, it may reach a height of 20' or more. It has light-brown bark and square stemmed twigs. The wood is very heavy, hard, dark-brown to black with light-brown sapwood and is sometimes used as a substitute for true boxwood in the making of engraver's blocks.

The leaves are usually opposite, simple, light green in color with entire or slightly crenate margins and 1 to 2" long. They are oblong-ovate or elliptical in shape with short petioles, a wedge-shaped base and rounded or blunt-pointed apex occasionally minutely notched.

The plants are unisexual with male and female flowers occuring on separate trees. The flowers are white with 5 spreading petals on long stalks in few-flowered terminal clusters. The fruits are dark blue to black, ovoid, 1 or 2-seeded, ¼" long drupes.

It was photographed near Fort Taylor in Key West in July, on Lower Matecumbe Key in flower in October and on Sugarloaf Key in fruit in March.

SCIENTIFIC SYNONYM: *G. grisebachii* Sarg.

Hamelia patens Jacq. RUBIACEAE
Common Names: **Firebush, Scarlet-bush.**

This evergreen shrub or small tree (10 to 12' high) is native to South Florida and the Keys, the West Indies and Tropical America. The leaves occur in whorls of 3 to 7 on red stalks. They are elliptic, 3 to 6" long, pointed, hairy, dull, dark green and at times purplish or red tinted.

The flowers are orange, red or orange-red, tubular with a slight flare, 5-lobed and 1 to 1½" long. They grow in tassel-like terminal or axillary clusters about 5" across. The fruits are dark red, purple or black, oval berries about ¼" in diameter, reportedly edible but quite seedy—a favorite, however, of many wild birds. Hummingbirds are attracted to the nectar in the flower tubes. It is also an excellent butterfly plant.

This plant was photographed on Key Largo in August.

SCIENTIFIC SYNONYM: *H. erecta* Jacq.

Hibiscus poeppigii (K. Spreng.) Garcke   MALVACEAE
Common Name: **Wild Hibiscus.**

This slender shrub at times reaches a height up to 6'. It is found in South Florida, Mexico, the West Indies and the Florida Keys.

The leaves are 1 to 3" long and are covered with fine hairs on the underneath side. They are generally deltoid-ovate in shape with irregularly tooth and 3-lobed margins. The flowers are up to 1" long and have bright red corollas. They are cylindrical, nodding and usually solitary in the upper axils. When fully mature, they do not completely open but remain in a half-closed position. The seed pod is a pubescent capsule ¼ to ½" long.

The specimen shown here was photographed on the edge of a hammock on Windley Key in August in flower.

SCIENTIFIC SYNONYM: *H. pilosus* (Sw) Fawc, & Rendle.

Hibiscus tiliaceus L. MALVACEAE
Common Names: **Mahoe, Tree Hibiscus, Sea Hibiscus.**

Considered by some to be native to the Florida Keys and South Florida and by others to be naturalized, this shrub is certainly well established and grows wild throughout the Keys. It occurs as a thicket-forming shrub or a densely crowned tree 20 to 30' high. The dark green leaves are heart-shaped, nearly round with a short tip and 4 to 6" wide. They are evergreen, leathery and edible when young.

The flowers are cupped, 5-petaled, bright yellow in the morning, turning pink and finally maroon by late afternoon. They are edible cooked as a vegetable or battered and fried. The fruits are long pointed capsules about ¾" wide containing many small, dark-brown seeds.

Specimens shown here are in Key West and Sugarloaf Key. They were photographed in January.

SCIENTIFIC SYNONYM: *Pariti tiliaceum* (L.) Britton.

Hippomane mancinella L.   EUPHORBIACEAE
Common Names: **Manchineel, Manzanillo, Poison-guava.**

This tree or shrub is native to the Florida Keys, West Indies and Central America. The champion in the National Register is in Watson Hammock on Big Pine Key—circumference of the trunk —3'11", height—39', spread—34'. It has light gray bark and wide spreading branches. The leaves are light green on long stems growing on short thick twigs. They are oval-shaped with pointed tips, deciduous, alternate, 2 to 4" long and minutely toothed along the margins. The flowers are small, green and yellow on 6" spikes followed by crab-apple shaped fruit 1½ to 2" across with a woody core containing several ¼" brown seeds.

Considered dangerously poisonous by most knowledgeable people, it has been eradicated to the point it is now considered "Threatened." The milky sap causes severe blistering and painful swelling on some people. Even a small amount in the eye can cause temporary blindness. The fruit is toxic when taken internally, possibly causing ulceration of the intestinal tract.

It was photographed in flower on Sugarloaf Key in September.

DESIGNATED STATUS: Threatened.

Hypelate trifoliata Sw. SAPINDACEAE
Common Names: **White Ironwood, Hypelate, Inkwood.**

This shrub or small tree, reaching 30' in height, is native to South Florida, the Keys and the West Indies. It is evergreen with alternate trifoliate leaves 3 to 5" long. The leaflets are 1 to 2½" long, dark green above, lighter below with close parallel veins. The leaves are persistent, leathery, wedge-shaped, slightly notched at the apex with thickened curled-under margins.

The flowers are small (⅛" diameter) with 5 pale-green petals growing on axillary or subterminal panicles. The fruits are black oval-shaped drupes about ⅜" in diameter. They have sweet, thin flesh covering a single oval seed in a thick brittle stone.

The very hard and heavy wood is used for fence posts, marine construction and tool handles. It is dark brown and close-grained.

The flowers were photographed on Big Pine Key in June, the fruit on Key Largo in August.

DESIGNATED STATUS: Threatened.

Ilex cassine L.   AQUIFOLIACEAE

Common Names: **Dahoon Holly, Dahoon, Cassena, Yaupon.**

This small upright tree, 20 to 30' high, is native to the Southeastern U.S.A. Coastal Plain, Cuba, the Bahamas and the Florida Keys.

The leaves are evergreen, alternate, leathery, elliptic to oblong-obovate, dark green, 2 to 4" long, smooth above, pubescent below and may have entire or lightly serrate margins above midpoint.

The flowers are dioecious and more numerous on male plants. Both sexes have 4 white petals, 4 sepals, 4 stamens and are about ¼" across. The fruit is a globose, red or orange-yellow drupe, ¼ to ½" in diameter, containing 1 to 4 seeds.

The flowers were photographed on Cudjoe Key in February and the fruit on Plantation Key in September.

Jacquinia keyensis Mez — THEOPHRASTACEAE

Common Names: **Joewood, Cudjoe-wood, Sea Myrtle, Barbasco.**

The Joewood is a very attractive small tree with a large smooth gray trunk. It is native to South Florida, the Keys and the Bahamas. The leaves are evergreen, paddle-shaped, stiff and leathery with curling-under edges and obscure veins. They are yellow-green in color, 1 to 3" long, rounded at the tip or shallowly notched and concentrated at branch terminals.

The flowers are ivory or pale yellow in color with 5-lobed calyces. They have 5-lobed, funnel-shaped corollas on terminal or axillary racemes. They are very fragrant. The fruit is a nearly globular berry about ⅓" in diameter turning yellow or orange-red when ripe and containing many small light-brown seeds.

The specimen shown here was photographed in flower in July · on Sugarloaf Key and is 17' tall and 22" in trunk circumference.

Krugiodendron ferreum (Vahl) Urb.   RHAMNACEAE
Common Names: **Black Ironwood, Leadwood.**

Native to Southeast Florida and the Keys, this tree has the heaviest wood in the U.S.A. It has a specific gravity of 1.3 and weighs 81 pounds per cubic foot. It grows to a height of 20 to 30' in hammocks near tidewater. It has gray bark with woody ridges. The leaves are thin, flat, glossy and medium to dark green when mature. They are broadly oval, opposite, 1½ to 2½" long, blunt pointed, sometimes slightly notched, evergreen and persistent for 2 or 3 years.

The flowers are small, yellowish-green with no corollas on axillary clusters. The fruit is a single-seeded, black, spherical drupe tipped with the base of the style. The black, thin flesh is juicy, sweet and edible raw.

The specimen shown here was photographed on Cudjoe Key in flower in June and is 18' high.

SCIENTIFIC SYNONYM: *Rhamnium ferreum* (Vahl) Sarg.

Laguncularia racemosa (L.) Gaertn.f.   COMBRETACEAE
Common Names: **White Mangrove, White Buttonwood.**

Grouped with the Red and Black Mangroves, this tree is very important in the formation of shoreline and the prevention of its erosion. This tree grows to the landward side of the other two and like the others is native to Florida and the Keys, the Bahamas and the West Indies. The leaves are light green above and below. They are thick, succulent, evergreen, oval, rounded at base and tip, opposite, persistent, sometimes notched at the top and 1 to 3" long.

The tiny, fragrant, velvety flowers, with 5 greenish-white petals, grow in clusters on terminal and axillary spikes. The fruits are greenish brown, 10-ribbed, single-seeded, obovoid, about ½" long. The dark-red seed may germinate while still on the tree.

The above flowers were photographed on Sugarloaf Key in May.

Lantana camara L. VERBENACEAE

Common Names: **Lantana, Yellow-sage, Shrub Verbena.**

This colorful shrub, considered by some to have become naturalized in Florida, is found growing wild throughout the Keys. It has been observed in the Keys up to 6' tall. It has square stems, usually hairy and prickly. It is a fast grower and spreads readily. The leaves are rough and up to 5" long, have toothed edges, are oval-shaped tapering to a point, are strongly aromatic and are poisonous to grazing animals.

The flowers are white, bright yellow and pink turning orange and red. They are small in flat clusters 1 to 1½" wide on stems from the leaf axils and they bloom all year. The fruit is a globular drupe, green turning dark blue. The green fruits are reportedly fatally poisonous when eaten in quantity, whereas the ripe fruits are also reportedly eaten by natives.

Some consider the plant to be a spreading pestiferous weed very hard to get rid of once it becomes established. These plants were photographed in Key West and Sugarloaf Key.

SCIENTIFIC SYNONYM: *L. aculeata* L.

Lantana involucrata L. VERBENACEAE
Common Names: **Wild Lantana, Wild Sage.**

The Wild Lantana is native to Southern Florida, the Keys, the West Indies and Tropical America. This shrub normally grows in thickets up to 5' tall along roads, trails and on the edges of the hammocks. It has pale yellow bark, is heavily branched and has no spines. The leaves are light green, rough in texture, with toothed edges, oval in shape and have a spicy aroma when crushed.

The flowers, very small, grow in heads or clusters about 1" across; they are white to lavender and fragrant. They bloom in the Keys off and on all year. The fruit is a small, purple drupe about $\frac{1}{8}$" in diameter.

The specimens shown here were photographed on Geiger Key in January.

SCIENTIFIC SYNONYM: *L. odorata* L.

Lasiacis divaricata (L.) A. Hitchc.   POACEAE
Common Name: **Wild Bamboo.**

This shrub (actually a Graminoid) is native to South Florida, the Keys, the Bahamas, the West Indies and South America. The culms or stems are generally smooth throughout. The main stem is woody, viny and may reach a length of 10 to 12'.

The blades are narrowly lanceolate, 2 to 8" long and ⅛ to ¼" wide. The few flowered panicles are terminal.

These specimens were photographed on Stock Island and Marathon in April.

Leucaena leucocephala (Lam.) de Wit   FABACEAE
Common Names: **Lead Tree, Jumbie Bean, White Popinac.**

This unarmed shrub, to 15', or a small tree, to 30', grows in the hammocks on the Keys. It is native to the West Indies and Tropical America; it is generally considered to be naturalized in the Keys and South Florida. The feathery bipinnate leaves with 4 to 8 pairs of dark green pinnae are very attractive. The leaflets are ½ to 1" long and slender-elliptic in shape.

The flowers are small, white, in fluffy globular clusters, from 1 to 1½" in diameter and grow in axillary spikes. The fruit is a flat, narrow, reddish brown pod, 6 to 8" in length, which contains 16 to 20 flat hard brown seeds. As the pods ripen they split open spilling the seeds on the ground where they sprout and frequently develop a dense thicket.

The young seed pods and the new shoots have reportedly been cooked and eaten as a vegetable and the ripe seeds roasted and used as a substitute for coffee.

The specimen photographed here is on Sugarloaf Key and was photographed in January.

SCIENTIFIC SYNONYMS: ***L. glauca*** (L.) Benth., ***Mimosa glauca*** L.

Lycium carolinianum Walt.   SOLANACEAE
Common Name: **Christmas Berry.**

This is a low, sprawling shrub or small tree (6 to 7' tall) with many curving, spiny branches. The leaves are dull gray-green, thick, succulent, ½" long, narrowly ovate, smooth with entire margins.

The flowers are white to pale lilac in color, 5-petaled, and about ¼" long. They grow on axillary stems either singly or in clusters. The fruit is a bright red berry, egg-shaped and about ½" across.

The plant grows on sand dunes, salt-pond edges, shell mounds or hammocks.

It was photographed in January with flowers on Little Torch Key, and in February on Geiger Key with fruit.

Lysiloma latisiliquum (L.) Benth. FABACEAE
Common Names: **Wild Tamarind, Bahama Lysiloma.**

Native to South Florida and the Keys, the Bahamas and West Indies, this is a medium size tree 40 to 60' tall. It has large horizontal branches forming a wide head making it an excellent shade tree. The leaves are feathery, twice-pinnate with many ½" long light-green leaflets oblong in shape.

The flowers are small, greenish-white in fuzzy, globular clusters, ½ to 1" across on axillary stalks. The fruit is a thin, flat pod, 3 to 6" long, reddish-brown when ripe, containing 8 to 10 hard, oval, brown, shiny seeds about ½" long. The pods usually hang on the tree until the next year.

The tree is salt tolerant, structurally strong enough to endure strong wind, does well in most soils and has no known serious pests or diseases. It grows rapidly from seed or seedlings, but because of its potential size, it is not recommended for small lots. The hard durable wood is considered valuable for boat timbers.

These pictures were taken on Plantation and Sugarloaf Keys. The flowers were photographed in March.

SCIENTIFIC SYNONYM: *L. bahamensis* Benth.

Mallotonia gnaphalodes (L.) Britt.   BORAGINACEAE
Common Names: **Sea Lavender, Beach Heliotrope, Bay Lavender.**

This very effective shoreline erosion control plant is native to South Florida, the West Indies, the Bahamas, Mexico and the Florida Keys. It is a sprawling, clump-forming, hairy, much-branched shrub 2 to 6' tall and frequently spreading over an area 20' or more in diameter.

The leaves are narrow, whitish to gray, pubescent and 2 to 4" long in dense terminal whorls. They are linear to spatulate, alternate, thick, with entire margins and persistent. The flowers are small, with white 5-lobed corollas, in clustered, 1-sided, curved terminal spikes. The fruits are nearly round, up to $\frac{3}{16}$" wide, black, ovoid drupes containing 2 seeds.

So much of this plant has been destroyed by thoughtless beach construction, it is now on the "Threatened List."

The plant, in flower, was photographed on Geiger Key in May.

SCIENTIFIC SYNONYM: *Tournefortia gnaphalodes* (L.) R. Brown.
DESIGNATED STATUS: Threatened.

Manilkara bahamensis (Bak.)
H. J. Lam & Meeuse   SAPOTACEAE
Common Names: **Wild Dilly, Wild Sapodilla.**

This evergreen shrub or tree may reach a height of 30 to 40'. It is native to South Florida, the Keys, the Bahamas and the West Indies. The wood is very heavy, hard, strong, close-grained and rich dark brown in color. It has milky sap, reddish-brown bark and a compact rounded crown.

The leaves are dull dark-green, leathery, alternate, persistent, and grow in twig-end clusters. They are oblong or elliptic in shape, notched at the apex, 2 to 4" long and covered with brownish fuzz underneath.

The flowers are light-yellow, 6-lobed, about ½" wide and occur in axillary, hairy-stemmed drooping clusters. The fruits are brown berries with spongy brownish flesh, milky juice, thick, scurfy skins, globular in shape and about 1½" in diameter. Edible raw, they are best when the juice has dried. There are 1 or more brownish-black flat seeds.

The specimens shown here are in hardwood hammocks on Sugarloaf Key and were photographed in May.

SCIENTIFIC SYNONYMS: *M. jaimiqui* (Wright) Dubard, *Achras emarginata* (L.) Little, *Mimusops emarginata* (L.) Britton.

Manilkara zapota (L.) Royen   SAPOTACEAE
Common Names: **Sapodilla, Naseberry, Chicle-gum Tree.**

Native to Central America, this tree has become naturalized in South Florida and the Florida Keys. It is a large tree. One was measured in Key West—51' high, 12'8" trunk circumference and 69' crown diameter.

The leaves are stiff, dark green, 2 to 5' long, elliptic and pointed in shape, of coarse texture and grow in rosette clusters. The flowers are about ½" across, white or cream colored and grow in leaf axils toward the end of branches.

The fruits are globular or oblate, 2 to 4" in diameter, light or dark-brown in color with a sandy surface. They contain a yellow-brown, greenish or reddish brown, juicy and very sweet pulp which is edible fresh, frozen or in preserves. The seeds are hard, shiny, black and often with small barbs which are dangerous if swallowed. The sticky white latex of the sap and unripe fruit yields chicle, used in making chewing gum.

The plant was photographed in Key West and Sugarloaf Key in May in flower.

SCIENTIFIC SYNONYMS: *M. zapotilla* (Jacq.) Gilly, *Achras zaputa* L.

Mastichodendron foetidissimum (Jacq.) SAPOTACEAE
H. J. Lam
Common Names: **Mastic, False Mastic, Jungle Plum, Wild Olive.**

This tree is a tropical hardwood found in hammocks of South Florida, the Keys, the Bahamas and the West Indies. It reaches a height of 50 to 70'. Its leaves and bark have a disagreeable odor. The heartwood is heavy, strong and bright orange in color. The sapwood is thick and yellow. It is used in cabinet work and boat construction.

The leaves are glossy, yellow-green, with wavy margins and pale midribs. They are 2 to 8" long, alternate, blunt-pointed, evergreen, elliptic in shape and clustered toward the end of the twigs. The flowers are light yellow, less than 1" in diameter and occur in clusters on orange-colored stems from the leaf axils. The fruit is yellow with a tough skin, has thick, white, juicy flesh; it is edible but subacid and somewhat bitter. The single seed is nearly round and about ½" long.

The tree was photographed with fruit in February on Key Largo.

Maytenus phyllanthoides Benth. CELASTRACEAE
Common Names: **Mayten, Gutta-percha Mayten, Florida Mayten.**

This shrub or small tree (18 to 20') is native to South Florida and the Keys. It has thin gray or gray-brown smooth bark and is very densely leaved, from which a substitute for Gutta Percha is obtained.

The leaves are thick, brittle, flat, green above and paler below, oblong or elliptical in shape, 1 to 1½" long and about half as wide. They are alternate, leathery, deciduous but persistent, with entire or wavy margins and have rounded or minutely notched tips.

The flowers are small, greenish-white, solitary or in few-flowered axillary clusters. The fruits are bright red, 4-angled, egg-shaped capsules about ¼" long with short spines at the tips. When the fruits mature, they split open to the base revealing 2 to 4 seeds enclosed in open scarlet arils.

This specimen was photographed in a hardwood hammock on Big Pine Key in February in flower and in June with fruit.

Metopium toxiferum (L.) Krug & Urb. ANACARDIACEAE
Common Names: **Poisonwood, Hog Gum, Florida Poisontree, Coral Sumac, Doctor Gum.**

The poisonwood tree is native to South Florida, the Keys, the Bahamas and the West Indies. It is a beautiful tree and quite abundant throughout the Keys. It belongs to the same family as the Poison Ivy, its poison sap causing dermatitis for some people. Its fruit is a choice food of the rare White Crowned Pigeons that inhabit the Keys.

The evergreen tree may reach a height of 30' or more. It has reddish-brown or gray bark, blackening when broken and covered with the sticky sap. The persistent and alternate leaves are pinnately compound with 3 to 7 but usually 5 dark-green, glossy, wedge-shaped leaflets, 3 to 4" long.

The small creamy-white flowers occur in panicled clusters in the leaf axils at the tips of the twigs. They are unisexual with the male and female flowers on separate trees. The fruits are oval in shape, about ½" long, dull orange in color, in loose dangling clusters and contain single ¼" brown seeds.

The specimens were photographed on Key Largo, Key West and Sugarloaf Keys in August.

Morinda royoc L.   RUBIACEAE

Common Names: **Yellow Root, Mouse's Pineapple, Morinda, Cheese Shrub.**

Native to South Florida and the Keys, this plant grows wild like a weed, up to 10', either as a vine or a shrub with reclining stems. The leaves are 2 to 4" long, obovate in shape, about 1¼" wide, pale green and lighter below, with prominent veins.

The flowers are small—¼" across, white, star-shaped with 5 or 6 petals, in compact heads on short stems out of the leaf axils. The fruits in the flower clusters fuse together forming a compound, globular drupe. They are green, turning yellow as they ripen, with a bumpy surface and many eyes. They are fleshy, about an inch in diameter and have an old cheese odor.

The plants pictured here are on Big Pine and Sugarloaf Keys. The flowers were photographed in May.

Myrica cerifera L.   MYRICACEAE

Common Names: **Wax-myrtle, Southern Wax-myrtle, Southern Bayberry.**

This shrub or tree is native to the Eastern U.S.A. Coast, the Florida Keys and the West Indies. It can form thick shrubby thickets or single trees to 30 or 40' in height. The bark is smooth and light gray. The wood is light, soft, brittle and dark-brown in color.

The leaves are persistent, elliptic-lanceolate or wedge-shaped, rounded or pointed, dull, thin, leathery, aromatic when crushed, entire or irregularly toothed, deep green above and paler below, alternate, and 2 to 4" long. They are used to flavor meats and broths. The flowers are small, yellowish-green in axillary clusters. Male and female flowers are on different plants. The fruits are globular, ⅛" in diameter in small clusters. They are coated with a bluish wax that is used to make fragrant candles. The fruits are also eaten by many birds.

This plant is found growing in hammocks, alongside sink holes, marshes and swamps. The plants shown here are on Big Pine Key, photographed in November.

SCIENTIFIC SYNONYM: *Cerothamnus ceriferus* (L.) Small.

Myrsine floridana A. DC. MYRSINACEAE
Common Names: **Myrsine, Rapanea.**

This shrub or small tree, occasionally reaching a height of 18 to 20', is native to South Florida, the Keys, the West Indies and Tropical America. It has thick branches and grayish, thin and usually smooth bark. The leaves are bright green on thick short stems clustered near the ends of the branches. They are oblong to elliptical in shape, alternate, blunt-pointed and usually with entire recurved margins. They are persistent, have a lustrous upper surface and are 2 to 5" long. The Mikasuki Indians supposedly mixed the leaves with their smoking tobacco.

The flowers are tiny, white with thin purple stripes, unisexual or polygamous, in 3 to 10-flower clusters in the leaf axils or above the leaf scars along the branches. The fruits are dark-blue or nearly black, dry or fleshy drupes containing a single bony seed.

This plant was photographed on Sugarloaf Key in December.

SCIENTIFIC SYNONYMS: *M. guianensis* (aubl.) Kuntze, *Rapanea guianensis* Aubl., *Rapanea punctata* (Lam.) Lundell.

Nectandra coriacea (Sw.) Griseb.   LAURACEAE

Common Names: **Lancewood, Florida Nectandra, Jamaica Nectandra.**

This plant appears in the coastal woodlands and hammocks as a densely branched shrub or a small tree 25 to 30' tall. It is native to South Florida, the Keys, the West Indies and Tropical America. It has reddish bark and round deposits of cork.

The smooth green leaves have yellow midribs, short stems and are from 2½ to 5" long. They are elliptical or narrowly oval in shape, aromatic, evergreen, alternate and have entire margins. The flowers are bisexual, have 6 creamy white sepals, are from ¼ to ½" across and appear in axillary or terminal panicles.

The fruit is a dark-blue or black nearly spherical drupe with a red or yellow cup and about ½" long.

The above photograph was taken on No Name Key in June.

SCIENTIFIC SYNONYM: *Ocotea catesbyana* (Michx.) Sarg.

Opuntia spinosissima (Martyn) Mill.   CACTACEAE
Common Name: **Semaphore Cactus.**

This plant, native to the Florida Keys, is on the official State of Florida list of "Threatened Plants" and is proposed for inclusion on the United States list. Several small colonies were found on Little Torch Key. It is a succulent shrub with spiny-fleshy stems or joints reaching a height of 6'. The branches and attached pads generally lie in a flat plane—thus the common name "Semaphore Cactus."

The pads are elliptic, curved, up to 10" long, 3" wide and have many areoles. The spines are salmon colored, turning gray, up to 1" long and usually occur 5 to 9 per areole.

The flowers are bright red with green sepals up to ⅝" wide. The fruit is a yellow obovoid berry, 1 to 2" long with few seeds.

It was photographed in flower on Little Torch Key in January.

SCIENTIFIC SYNONYM: *Consolea corallicola* Small.
DESIGNATED STATUS: Threatened.

Opuntia stricta Haw.   CACTACEAE

Common Names: **Prickly Pear, Dillen's Prickly Pear.**

This succulent shrub is native to Southern U.S.A., the West Indies, Mexico and is quite common throughout the Florida Keys. It has fleshy, much-branched stems composed of obovate, spatulate or elliptic flat joints 4 to 10″ long and 3 to 5″ wide. These joints or pads are covered with conspicuous stiff sharp spines in clusters of 2 or more.

    The flowers are solitary, with bright-yellow or rose-tinted petals and are 2 to 4″ across. The author has observed them blooming during every month of the year in the Keys. The fruit is a pear-shaped berry, red to purple in color and 2 to 3″ long. It contains a sweet, juicy and purplish pulp that can be eaten raw or made into a jelly, jam or syrup. It is covered with tufts of very small spines or glocids which can be wiped off with a handful of dried grass before handling. The fruit is a choice food for many birds and the joints can be fed to cattle after the spines are burned off.

    The flowers shown here were photographed on Sugarloaf Key in March.

SCIENTIFIC SYNONYMS: *O. zebrina* Small, *O. keyensis* Britt., *O. dillenii* Haw.
DESIGNATED STATUS: Threatened.

Persea borbonea (L.) K. Spreng.　　　　　LAURACEAE
var. *pubescens* (Pursh.) Little

Common Names: **Swamp-bay, Swamp Red Bay, Red Bay, Bull-bay, Sweet-bay, 'Tisswood, Red-bay Persea.**

This hardwood may occasionally reach a height of 60 to 70' and is native to Eastern U.S.A., the Keys and the Bahamas. It has reddish-brown deeply furrowed bark. The newer branches are ruddy and smooth. The wood has been used in cabinetry and interior finishing.

The bright-green, leathery, alternate leaves are glossy above and dull below. They are evergreen, toothless, elliptic to oblong-lanceolate in shape, aromatic when crushed and from 3 to 6" long. They are sometimes pointed at both ends and have orange colored midribs. They are used fresh or dried to flavor meats, poultry, soups and stews. The Mikasukis used the leaves to make a tea.

The flowers are very small in axillary or terminal clusters. They have a bell-shaped, 6-lobed, pale-yellow or cream calyx and no petals. The fruits are deep-blue to violet-black ovoid berries containing 1 large round seed. They have a leathery skin and practically no pulp.

This tree was photographed in a hammock on Big Pine Key in July.

SCIENTIFIC SYNONYMS: *P. littoralis* Small, *P. palustris* (Raf.) Sarg., **Tomala borbonea** (L.) Raf., **Tomala pubescens** (Pursh) Sarg.

Pinus elliottii Engelm.
var. *densa* Little & Dorman
PINACEAE
Common Names: **Slash Pine, South Florida Slash Pine, Rock Pine.**

A very valuable tree for lumber, pulpwood, naval stores and resin, this tree is native to South Florida, many of the Keys, the Bahamas and the West Indies. It grows to a height of 50 to 75' in the Keys. The tree has scaly and deeply furrowed bark. Its trunk is generally unbranched until it reaches a dense symmetrical crown. The wood is very hard, durable and coarse-grained.

The needle-like leaves are 8 to 12" long, dark green, glossy, persistent until the second year and stiff in clusters of 2 or 3. Male and female flowers occur on the same tree. The male flowers are dark purple in many-flowered clusters, whereas the female flowers are generally solitary on long stalks and are pinkish in color.

The seed cones are ovate in shape, 2 to 6" long and open to release the seeds. Each scale of the cone has a short recurved spine. The ¼"-long, black seeds are attached to wings which help in their distribution when blown by the wind.

The specimens shown here were photographed on Sugarloaf Key in January.

SCIENTIFIC SYNONYM: *P. caribaea* Morelet.

Piscidia piscipula (L.) Sarg.   FABACEAE

Common Names: **Jamaica Dogwood, Fishfuddle Tree, Fish-poison Tree.**

Native to South Florida, the Keys, West Indies, Bahamas and Tropical America, this hardwood tree reaches a height of 50' or more. It has gray-green scaly bark which was used along with the roots, twigs, and leaves to stupefy fish and aid in their capture. This practice is now in violation of the Florida State law.

The leaves are pinnately compound with 5 to 7 grayish-green oval leaflets 1½ to 4″ long. They are deciduous, dark-green above and lighter below, alternate and have entire wavy margins. The flowers are pea-like with white and lavender or pink 5-lobed petals, borne in elongated clusters before the leaves appear. They are about ¾″ long and a favorite of bees. The fruits are 3 to 4″ long light-brown pods with 4 papery wings containing red-brown seeds.

The photographs were taken on Sugarloaf Key in May. The wood is very hard, heavy, durable, yellow-brown and valued for boat building, wood carving, fence posts and charcoal.

SCIENTIFIC SYNONYMS: *P. erythrina* L., *Ichthyomethia piscipula* (L.) Hitch.

Pisonia aculeata L. NYCTAGINACEAE
Common Names: **Devil's Claws, Cockspur, Pull-and-Hold-Back.**

Native to South Florida, Tropical America, the West Indies and the Florida Keys, this plant usually takes the form of a sprawling tree-like woody vine with dark purplish-black bark. The stems are armed with very sharp, paired, curved or hooked axillary spines. The leaves are ovate, oval or elliptical in shape and about 1 to 3" long. They are soft, light green and dull.

The yellow-green flowers have no petals. The multi-flowered axillary and terminal inflorescences are 1 to 2½" long. The fruits are dry, angled and covered with short soft hairs.

These plants are usually found in hammocks and on disturbed sites. Considered a weed in some areas, the spines can make the unwary very uncomfortable.

The plant was photographed on Key Largo in September.

Pisonia rotundata Griseb.   NYCTAGINACEAE
Common Name: **Pisonia**.

This shrub or small tree is native to the Lower Keys, the Bahamas and Cuba. It has ashy gray bark marked with large leaf scars. Its stems are unarmed. As a tree it may reach a height of 10 to 15'.

The leaves are dull-green with depressed veins. They are opposite, blunt-pointed, stiff, obovate or elliptic-ovate in shape with slightly curled under margins and from 1 to 4" long.

It has small green and white flowers with no petals that grow in densely flowered flat-topped inflorescences. The fruits are very small, angled and dry.

The illustrative specimen was photographed on Sugarloaf Key. The flower picture was made in April.

SCIENTIFIC SYNONYM: ***Torrubia rotundata*** (Griseb.) Sudw.

Pithecellobium guadalupense (Pers.) Chapm.   FABACEAE
Common Names: **Blackbead, Guadeloupe Blackbead, Ram's Horn.**

This is a tropical hardwood shrub or small tree to 20', native to South Florida, the Keys, the Bahamas, Guadeloupe, Cuba and the Yucatan. It is rarely thorny and has compound leaves with leaflet stems longer than leaf stems. The leaflets are evergreen, leathery, elliptic to round-oval, notched or pointed tip and from ½ to 2½" long. Each leaf generally has 4 leaflets in pairs.

The very small fragrant flowers are cream or pink in color in stalked, downy, globular heads from ¾ to 1¼" in diameter. The brown seed pods, 2 to 4" long, open when ripe in a twisted, curved or coiled form, disclosing shiny black seeds with bright red arils.

The black seeds may be drilled and strung as beads; the red arils are edible and have a sweetish flavor.

The plants shown here were photographed in flower on Sugarloaf Key in March.

SCIENTIFIC SYNONYM: *P. keyense* Britt. ex Coker.

Pithecellobium unguis-cati (L.) Benth.　　　　　　FABACEAE
Common Names: **Catclaw, Cat's Claw, Catclaw Blackbead.**

This shrub or small tree may reach a height of 20 to 25'. It is native to Florida, the Keys, the Bahamas, the West Indies and Tropical America. It can be distinguished from the Blackbead *(Pithecellobium guadalupense)* by the leaf stems, which are longer than the leaflet stems. It is also more likely to have stipular spines at the leaf bases.

The leaves are bipinnate with 1 pair of bifoliate pinnae. The leaflets are bright green, persistent, broad oblong, rounded on the one side and wedge-shaped on the other. New growth leaves are pinkish. The flowers are globular heads in terminal clusters or solitary on axillary stems. The corollas are white, pinkish or pale yellow. Seed pods are reddish brown, coiled, constricted between seeds and from 2 to 4" long. They open in a very contorted shape to expose black seeds and red arils. The wood is very hard, heavy, close-grained, rich red varying to purple, with clear yellow sapwood.

The photographs were taken on Lower Matecumbe Key in May.

SCIENTIFIC SYNONYM: *Mimosa unguis-cati* L.

Pluchea odorata (L.) Cass.   ASTERACEAE
Common Name: **Marsh Fleabane.**

This shrub is native to Coastal Southeastern U.S.A. and the Florida Keys. It is very similar to ***Pluchea symphytifolia*** except the leaves are smaller, the plant does not usually grow as tall, and the leaves have short petioles. The leaves of ***Pluchea odorata*** have serrate margins and acute apexes.

The flowers are pink, lavender or purple in color and grow in large terminal clusters. The fruits are very small, black, thickly pubescent achenes.

This plant was photographed on Long Key in May.

Pluchea symphytifolia (Mill.) Gillis   ASTERACEAE
Common Name: **Bushy Fleabane.**

This roadside shrub is native to South Florida, Tropical America and the Florida Keys. It is much branched and usually grows to a height of 3 to 10′. It seems to thrive on filled soil or bulldozed areas in the margins of hammocks.

Its leaves are lanceolate-ovate to elliptic-ovate in shape, 3 to 6″ long and covered with fine gray hairs. The leaf margins are entire or obscurely serrate. They have definite petioles of ½″ in length or more.

The very small flowers have pale pink to lavender corollas and occur in dense terminal clusters, followed by dry, single-seeded fruits.

The illustrative plants were photographed in Tavernier and on Sugarloaf Key in March.

Pseudophoenix sargentii H. Wendl. ex Sarg.   ARECACEAE
Common Names: **Buccaneer Palm, False Phoenix, Hog Palm, Sargent's Cherry Palm.**

This palm tree is native to Florida, the Keys, the Bahamas, the West Indies, Cuba, and the Yucatan. It is very slow growing but may reach a height of 15 to 20'. Its fruit was used by early settlers to fatten hogs. The terminal bud was cooked like cabbage and the sap was used to make wine. It is considered "very rare" in the wild state in the Keys.

The trunk is smooth and glossy, light gray-green with prominent rings close together and bulges at the ground line. The leaves are V-shaped, pinnate, stiff, unarmed, blue-green, arching prominently and 3 to 6' long.

The greenish-yellow flowers appear on 3' stalks branching in wide clusters. The fruits are bright red, round or lobed and about ½" in diameter.

The specimen was photographed in flower on Sugarloaf Key in September.

DESIGNATED STATUS: Endangered.

Psidium guajava L.   MYRTACEAE
Common Name: **Guava.**

This tree or shrub is native to Central America but has become naturalized in South Florida and the Florida Keys. It may form a shrubby thicket or become a slender spreading tree 25 to 30' tall. The leaves are dull green with conspicuous veins, opposite, elliptic or oblong in shape, evergreen, rounded or narrowed at the tips and up to 6" long. The flowers have white petals and white and yellow stamens. They are 1 to 1½" wide and grow singly in the leaf axils.

The fruit is spherical in shape and has a persistent calyx. It ripens yellow and is 1½ to 2" in diameter on the wild tree. It is edible raw, cooked or juiced. The flavor varies from sour to sweet. It is used for jelly, desserts and drinks. The bark and roots have been used medicinally.

The plants were photographed on Big Pine and Sugarloaf Keys. The flower picture was taken in May.

Psidium longipes (Berg.) McVaugh   MYRTACEAE
Common Names: **Long-stalked Stopper, Trailing Eugenia.**

This plant is native to South Florida, the Keys and the Bahamas. It is a shrub or small tree from 1 to 12' high. It normally has a short trunk and many wiry trailing branches up to 3' long.

The leaves are glossy, light green above and paler below, ovate or oval in shape, ½ to 1½" long with reddish veins on the under side. The flowers are white or pink, approximately ½" wide, fragrant, with dense bundles of white stamens and usually appear solitary.

The fruit is red turning dark purple or black on exceptionally long stalks. They are globose, from ¼ to ½" in diameter and contain many small seeds. The ripe fruit is edible raw.

The photographs were taken on Cudjoe and Sugarloaf Keys. The flower picture was taken in May.

Psychotria ligustrifolia (Northrop) Millsp.   RUBIACEAE
Common Names: **Wild Coffee, Bahaman Wild Coffee.**

This shrub is native to South Florida, the West Indies and the Florida Keys. It has glabrous or nearly smooth stems and may become 6' or more tall. The leaves are 2 to 5" long, lanceolate to oblanceolate in shape tapering gradually to a pointed apex. Stipules are usually quite noticeable at the bases of the leaves.

The white flowers appear in stalked clusters in the upper leaf axils. The fruits are ellipsoidal drupes about ¼" long that become red when ripe.

This plant was photographed on Key Largo in September.

SCIENTIFIC SYNONYM: *P. bahamensis* Millsp.

Psychotria nervosa Sw. RUBIACEAE
Common Name: **Wild Coffee.**

This shrub is not a substitute for real coffee as the common name implies. It is not considered edible by humans but is a popular food for birds. It is native to South Florida, the Keys, the Bahamas and the West Indies. A small shrub, it at times reaches a height of 8 to 10'. The leaves are bright green and shiny. The leaf tissue bulges above the midrib and veins giving a wrinkled appearance. They are evergreen, opposite, elliptic in shape, pointed and from 2 to 6" long.

The small flowers are white, 4 to 5-lobed, tubular and recurved, ⅓" long and borne in terminal and axillary clusters. The fruits are bright red, oval drupes enclosing 2 hemispherical seeds very similar to coffee beans.

The plant in flower was photographed on Sugarloaf Key in May.
SCIENTIFIC SYNONYM: *P. undata*.

Psychotria punctata Vatke   RUBIACEAE
Common Name: **Wild Coffee.**

This branching shrub grows up to 8' or more in height with glabrous stems. It is native to Africa but has become naturalized in the Lower Keys. Its leaves are elliptic-obovate to ovate in shape, from 1¼ to 2½" long, rounded or short-pointed at the apex, with upper surfaces raised above the veins and spotted on the underneath side with nitrogenous nodules.

The small flowers, in rounded axillary clusters, have white corollas. The fruits are ovoid drupes, red or crimson in color and about ¼" in diameter.

The plants shown here were photographed on Stock Island in flower and fruit in June.

SCIENTIFIC SYNONYM: *P. bacteriophylla* Valet.

Quercus virginiana Mill. FABACEAE
Common Name: **Live Oak.**

This tree is found on the Coastal Plain of Southeastern U.S.A., Central America, Mexico and Cuba. It is very rare in the Florida Keys. The plant is an erect tree with dark bark and very hard wood. It is considered valuable for its timber.

It is evergreen in that its leaves are persistant and remain on the tree until the new leaves of the next season appear. They are elliptic or obovate in shape, 1 to 4" long, dark green and shiny above, paler and slightly toothed and maybe somewhat rolled back.

The plant is monoecious with male and female flowers on separate spikes. The male flowers are in hanging tassels whereas the female flowers are on short, erect spikes. The fruit is a 1-seeded acorn about 1" long and less than ½" wide with a cup covering approximately ¼ of the nut.

The plant shown here was photographed in a wild hammock on Upper Key Largo.

Randia aculeata L.   RUBIACEAE

Common Names: **Randia, Box-brier, White Indigo-berry, Indigo-berry.**

This shrub or small tree, up to 10' tall, is native to South Florida, the Keys, the Bahamas and the West Indies. The erect plant has opposite, horizontal branches and is used in the Virgin Islands as a Christmas Tree. The wood is very hard and heavy and is used in the Netherlands Antilles to make cooking tools and fishing rods. The leaves are light green, shiny, blunt pointed with a small, sharp tip. They are opposite, elliptic to nearly round in shape, thickened slightly and ½ to 2" long.

The flowers are white, with 5 petals, clustered in leaf axils or solitary along the stems. They are fragrant and about ½" across. The fruits are white or greenish-white, round or oval, containing several rounded flat seeds in a blue-black pulp, hence the common name "White Indigo-berry."

The specimen in flower was photographed on Sugarloaf Key in May.

Reynosia septentrionalis Urb.　　　　　　RHAMNACEAE
Common Names: **Darling Plum, Red Ironwood.**

This shrub or tree, reaching a height of 20' or more, is native to South Florida, the Keys, the Bahamas and the West Indies. Its wood is heavy, very hard, strong, close-grained, rich dark-brown in color with light sapwood. The bark is gray-brown and the evergreen leaves are 1 to 1½" long, leathery, opposite, variable in shape but usually oblong-oval or obovate and notched at the tip. They are thin, stiff, dark-green above and paler or light brown below. New growth is pinkish in color.

　　The flowers do not have petals, but 5 yellowish-green, ovate and sharp-pointed sepals growing in axillary clusters. The fruits are oval or round, dark-purple drupes with spine tips. The flesh is thin, very sweet and covers a round, rough and light-yellow seed stone. The flesh is edible raw or cooked.

　　It was photographed in flower on Big Pine Key in March and in fruit on Sugarloaf Key in June.

Rhizophora mangle L.   RHIZOPHORACEAE
Common Name: **Red Mangrove.**

One of the most valuable trees of the Island Tropics is the Red Mangrove whose adventitious roots contribute so much to the creation and preservation of low and marginal land along the shore. It is native to Florida, the Keys, the Bahamas, the West Indies and Tropical America. It usually is 20' or less in height but sometimes reaches 35' or more. It grows in shallow salt water standing on curved aerial prop roots, forming dense thickets along the water's edge.

The leaves are evergreen, leathery, 2 to 6" long, ovate to elliptic in shape, dark green above and paler below and persistent for 1 or 2 years. The flowers are pale yellow, less than 1" across, have 4 petals with 2 or 3 in axillary clusters and appear throughout the year. The conical 1" long brown fruit contains a seed that germinates and develops a 12" long seedling while still on the tree. This cigar-shaped sprout drops from the tree, floats vertically until it becomes snagged on the bottom, sends out roots and starts a new tree.

The hard, heavy, close-grained, durable wood is dark-brown streaked with light-brown. The bark is a source of tannic acid.

The photographs were taken on Big Pine and Geiger Keys in September.

DESIGNATED STATUS: Special Concern.

Ricinus communis L.   EUPHORBIACEAE
Common Names: **Castor Bean, Castor Oil Plant.**

Generally considered to be a native of the Old World tropics, this plant has become naturalized in South Florida and the Florida Keys. In fact, it has frequently been classified as a "Weed Tree" in some areas. It may reach a height of 15' or more in the Keys.

The large palmate leaves with 7 to 9 deeply cut lobes on long stems are green or purplish and coarse in texture. The flowers appear on long upright spikes at the ends of the branches. The fruit pods are burr-like, green, bluish or red, about 1" across, becoming dry and brown as they ripen. The dry pods split open to free white or tan seeds, mottled with dark-brown or black, from ¼ to ½" long. These seeds are poisonous and are a source for castor oil and lubricating oil.

The plants were photographed in Key West in flower in January and November.

Rivina humilis L.   PHYTOLACCACEAE
Common Names: **Rouge Plant, Bloodberry.**

This shrub is native to Southeastern U.S.A., Mexico, Central America, the West Indies and the Florida Keys. It grows as a vine or erect to 5'. The older specimens generally have a woody base. The leaves are alternate, simple, entire, lanceolate to ovate-elliptic, tapering to a point and up to 5" long.

The small flowers occur on terminal sub-erect spikes up to 3" in length. The flowers have 4 sepals, white to pink, no petals and 4 stamens. The fruit is a globose, red berry when ripe and about ⅛" in diameter.

It is usually found on the margins of hammocks in the Keys. The flower picture was taken on Geiger Key in October.

Roystonea elata (Bartr.) F. Harper ARECACEAE
Common Names: **Florida Royal Palm, Royal Palm.**

This palm tree is native to South Florida and Cuba. It has not been definitely accepted at the present time as a native of the Keys. It has a massive trunk, light gray and very smooth. This trunk is topped by another column known as the crown shaft, composed of clasping bases of petioles of the leaves and is bright green. Above this shaft is the crown of deep green arching pinnate leaves up to 12' in length. The leaflets are tapering and 2 to 3' long in 2 rows on either side of the rachis, in different planes making the leaf appear to be roundish.

The flower stalk is long and loose and the fruits are almost round, about ½" long and violet blue in color when ripe.

These trees are very rare in the wild state outside of the Everglades National Park and the Fakahatchee Strand State Preserve.

The tree shown here is on Sugarloaf Key and was photographed in a private garden in May.

SCIENTIFIC SYNONYMS: *R. floridana* O. F. Cook, *R. regia* Cook.
DESIGNATED STATUS: Endangered.

Sabal palmetto (Walt.) Lodd. ARECACEAE
ex J. A. & J. H. Schult.
Common Names: **Cabbage Palm, Palmetto Palm, Carolina Palm.**

This famous palm, the Florida State Tree, is native from North Carolina to Florida, the Keys, the Bahamas and the West Indies. It was a very important food plant for the Indians and the early settlers. The leaf bud or heart was eaten raw, used in a salad or cooked as cabbage. It was also dried and made into a meal for bread. The fruits were made into a syrup or eaten raw.

The tree is tall, growing to 60', frequently partly covered with plaited old leaf bases, which drop off, leaving the trunk relatively smooth and gray. The leaves are fan-shaped, 3 to 6' long, medium glossy green above and grayish green below, deeply divided on smooth 4 to 6' petioles. There are many threads in the sinuses. The leaves are costapalmate.

The flowers are greenish-white, fragrant, ¼" wide in large drooping clusters on 2 to 3' stalks. The fruit is globular, blackish-brown and about ½" in diameter.

The photographs were taken on Sugarloaf Key in October.

SCIENTIFIC SYNONYM: *S. jamesiana* Small.

Sapindus saponaria L. SAPINDACEAE
Common Names: **Soapberry, Wingleaf Soapberry.**

This tropical hardwood tree is native to the Bahamas, the West Indies, Tropical America, South Florida and the Florida Keys; it may reach a height of 20 to 30'. The sap has been used for soap and the fruits, rubbed between the hands, produce a light cleansing lather. The wood is hard, tough and light brown in color.

The pinnately compound leaves are medium green when mature and much lighter when young. There are generally 4 to 9 elliptical to lance-shaped, pointed leaflets that are toothless, 2 to 4" long, smooth and borne on winged rachis.

The flowers are small with 4 to 5 white petals and occur in many-flowered terminal or axillary clusters, followed by globular orange-colored juicy berries about ½" in diameter.

This plant was photographed on Key Largo in June and on Big Pine Key in flower in November.

Savia bahamensis Britt.   EUPHORBIACEAE
Common Name: **Maiden Bush.**

A shrub or small tree to 10' high, the Maiden Bush is native to Florida, the Keys, the Bahamas and the West Indies. It has smooth pale-gray bark. The leaves are ovate or abovate in shape, 1 to 2" long, blunt pointed with entire margins, short petioled, leathery, shining gray-green above and paler below. They have a light midrib and obscure veins.

The plant is dioecious. The flowers are small with 5 greenish petals and a 5-lobed calyx. The staminate flowers have 5 stamens and appear in dense axillary clusters. The pistulate flowers are in few-flowered clusters or solitary.

The fruit is a dry brown capsule containing seeds less than ¼" long.

The specimen shown here was photographed on Sugarloaf Key in July.

Scaevola plumieri (L.) Vahl GOODENIACEAE
Common Names: **Ink-berry, Beach-berry, Scaevola.**

This shrub is native to South Florida, the Keys, Bermuda and the West Indies. It has many drooping branches which frequently root as they touch the ground forming thick clumps. It reaches a height of 3 to 5'.

The leaves are thick, fleshy and from 1 to 2½" long on short, winged petioles. They are green, glossy and obovate or spatulate in shape with entire margins and are concentrated at the ends of the branches.

The white or pinkish-white flowers occur in small clusters among the terminal leaves. They have 5 or 6 lobes shaped like a hand or fan. The fruits are smooth glossy-black drupes from ¼ to ½" in diameter. They are juicy, bitter and enclose a 2-seeded woody stone.

These plants were photographed on Bahia Honda Key in May.

DESIGNATED STATUS: Endangered.

Schaefferia frutescens Jacq.   CELASTRACEAE
Common Names: **Florida Boxwood, Yellow-wood, Boxwood.**

This deciduous tree or shrub with slender and smooth stems occasionally reaches a height of 30' or more. As a shrub it usually is 10' or less. It has thin brown bark turning light gray as the branches become older and marked by persistent wart-like clusters of bud scales. The leaves are alternate, persistent, elliptical or oval in shape. They are shining bright yellow-green, with thickish rolled-under margins, 2 to 3" long and ½ to 1" wide.

The 4-petaled greenish flowers, ¼ to ½" across appear in axillary clusters. Small fleshy bright-red drupes about ¼" in diameter contain a bony stone with 2 seeds. The wood is heavy, close-grained, bright clear-yellow and is often used in wood carving.

This plant is native to South Florida and the Keys, Tropical America and the West Indies. It is usually found in hammocks and sandy soil near the tidewater.

This specimen was photographed on Key Largo in February.

Schinus terebinthifolia Raddi   ANACARDIACEAE
Common Names: **Brazilian Pepper, Florida Holly, Christmas Berry, Pepper Tree.**

This plant, frequently classified as a pestiferous weed, was imported from Brazil to South Florida in 1891 as an ornamental shrub. It has since become naturalized and grows abundantly throughout South Florida and the Keys, in many instances, crowding out more desirable native species. It is very difficult to control once it gets started.

   It grows as a tree or as a clump-forming shrub to a height of 25' or more. It is evergreen and the leaves are alternate and pinnately compound with 3 to 11 opposite lanceolate to elliptic leaflets to 3" long. The plant is dioecious with small white flowers in large axillary clusters usually grouped at the end of the branches. The fruit is a bright red spherical drupe about ⅛" in diameter in large clusters of 100 or more. They are juicy and very attractive to birds and some animals. Like its relative, Poison Ivy, it causes severe skin disorders for some people. When in flower it often causes respiratory discomfort.

   It was photographed in flower on Sugarloaf Key in October.

SCIENTIFIC SYNONYM: *Rhus terebinthifolia* Schlecht. & Cham.

Schoepfia chrysophylloides (A. Rich) Planch.   OLACACEAE
Common Names: **Graytwig, Gulf Graytwig, Schoepfia, Whitewood.**

This shrub or small tree is found in South Florida, the West Indies, the Bahamas, Tropical America and the Florida Keys. It is unarmed, with light furrowed bark, almost white on the twigs, and may reach a height of 10' or more. It has been known to grow as a parasite on other trees.

The leaves are alternate, with wavy edges, 1 to 2½" long, ovate-elliptic, entire, leathery, simple, usually pointed and narrowed to a short petiole. The sweet-scented flowers are red and orange-tinted in complex clusters in current-growth leaf axils. They have 4 or 5 petals that are fused below to form a tube with recurved lobes. The fruits are 1-seeded red drupes which turn black at maturity. They are ovoid in shape, up to ½" in diameter and have a reddish pulp covering a hard brittle stone.

It was photographed on Lignum Vitae Key in December.

SCIENTIFIC SYNONYMS: *S. schreberi* Gmel., *S. americana* Willd.

Serenoa repens (Bartr.) Small   ARECACEAE
Common Name: **Saw Palmetto.**

This shrub is native to Southeastern U.S.A. from the Florida Keys to South Carolina. It normally has a twisted recumbent trunk but at times it grows erect up to 10'. The reclining trunks and underground stems take root and cause the plant, at times, to spread into large colonies.

The leaves are fan-shaped, nearly circular, stiff, usually green to yellowish-green and without a midrib. They are deeply cut into many divisions and are usually about 3' in diameter. The petioles are slender, 4 to 5' long and armed with small curved sharp spines, hence the name "Saw Palmetto."

The small ivory-white flowers are fragrant, in 1 to 3' long plume-like clusters emerging from among the leaves. They are a favorite nectar source for honey bees. The fruits are dark blue or black when ripe, oval to oblong in shape, have edible light-brown pulp, about ½" in diameter, up to 1" in length, and contain oval light brown seeds.

The flowering plant was photographed on Sugarloaf Key in May.

Simarouba glauca DC.   SIMAROUBACEAE
Common Names: **Paradise Tree, Bitterwood.**

This tree is native to South Florida, the Keys, the Bahamas and the West Indies. It is a slender tree with reddish brown and gray bark; at times it reaches a height of 50'. It is a source for resins, oils and pharmaceutical compounds.

The leaves are pinnately compound, persistent and 6 to 16" long. There are 10 to 14 opposite or alternate leathery leaflets, 2 to 4" long, dark glossy-green above and light gray underneath. They are oval, oblong or obovate in shape and the new growth is reddish.

The cream-colored or yellow flowers are small, 4 or 5-petaled, occuring profusely in axillary or terminal clusters. The fruit is an oval 1" long drupe. It changes from red to purple or nearly black when ripe. It has firm, juicy, white, sweetish flesh and is edible but rather insipid. It contains 1 orange-brown rough seed about ¾" long.

The trees shown here were photographed on Plantation and Sugarloaf Keys in April.

Solanum bahamense L.   SOLANACEAE
Common Names: **Bahama Nightshade, Canker-berry.**

This shrub is native to the Florida Keys, South Florida and the West Indies. It is found throughout the keys growing in thickets or along the roadside on the edges of hammocks. The woody stems are hairy and erect with many smooth or spiny branches. It reaches a height of 2 to 6'. The leaves are lanceolate to oblong, rough textured and 2 to 5" long with smooth or wavy edges.

The flowers, in branched clusters, have 5 violet-blue (sometimes pale) petals flared star-like and about ½" across. They are very similar to the domestic potato. The fruits are brilliant-red berries, global in shape and from ¼ to ½" in diameter on drooping stems.

It was photographed on Big Pine Key in flower in February.

Solanum donianum Walpers   SOLANACEAE
Common Name: **Potato Tree.**

Native to South Florida, the Florida Keys and the West Indies, this shrub reaches a height of 5' or more. The leaves are simple, leathery with wavy margins and from 2 to 5" long.

    The flowers are white, about ½" wide and usually occur in upright clusters at the terminal ends of the branches. The fruits are red globular berries in clusters on erect stems.

    These shrubs are usually found along filled access roads on the edges of hammocks. The above specimen was photographed in such an area on Sugarloaf Key in flower in May.

Solanum erianthum D. Don   SOLANACEAE
Common Names: **Potato Tree, Mullein Nightshade.**

This plant belongs to the same family as the tomato, white potato, tobacco, eggplant, and the poisonous nightshade. It grows either as a shrub or a small tree and is native to South Florida and the Keys, the West Indies and Tropical America. The leaves, stems and flowers are covered with wooly hairs. The leaves are light green, oval or elliptic in shape and vary from 4 to 10" in length with slightly wavy margins. They are reportedly used in Mexico to clean dishes and relieve headaches.

The flowers have 5 star-shaped white petals and are ½ to ¾" in diameter. They appear in stalked flat-topped clusters near the leaf axils. The fruits are berry-like, globular, ¼ to ½" in diameter, yellow in color and contain many small seeds.

The plant is found throughout the Keys in thickets and on the edges of hammocks. This specimen was photographed on Sugarloaf Key in flower in July.

SCIENTIFIC SYNONYM: *S. verbascifolium* Small.

Sophora tomentosa L. FABACEAE
Common Names: **Necklace-pod, Hairy Sophora, Silver Bush.**

This shrub is native to Florida, the Keys and the West Indies. Although usually 3 to 4' high, it can at times reach a height of 10 to 15'. Its bark is yellowish-brown. The odd-pinnate compound leaves are up to 12" long and have 11 to 21 opposite, elliptic or oval recurved leaflets. They are ½ to 2¼" long, leathery, dark green, with entire margins and prominent midveins. They are covered with silky-silvery fuzz when young, hence the name *"tomentosa."*

The flowers are yellow, pea-shaped, about 1" across, growing on terminal spikes 4 to 16" long and opening progressively from the base to the tip. The seed pod, 2 to 8" long and about ½" in diameter, is velvety-brown and constricted between the 2 to 9 seeds, making it look like a necklace.

The specimen shown here was photographed on Sugarloaf Key in October.

Stachytarpheta jamaicensis (L.) Vahl   VERBENACEAE
Common Names: **Blue Porterweed, Rat's Tail.**

This shrub is native to South Florida, the Keys, the West Indies and Central America. Although it usually has spreading or decumbent branches close to the ground, it does at times reach a height of 4 to 5'. Its leaves are 1 to 4" long in an oblong or ovate-lanceolate shape with toothed margins.

The flowers appear in quill-like spikes 4 to 6" long. The corollas are dark blue to violet, about ⅜" long and about ¼" wide. They open early in the morning and close around noon on hot days.

It is reported that the natives of Central America make a foaming tea from the leaves.

This plant was photographed on Cudjoe Key in May.

SCIENTIFIC SYNONYM: *Valerianoides jamaicensis* (L.) Kuntze.

Strumpfia maritima Jacq.   RUBIACEAE
Common Names: **Pride-of-Big-Pine, Strumpfia, Snowbank.**

This low shrub is on Florida's "Endangered List." It is native to the Bahamas, the West Indies, the Yucatan and the Florida Keys. It usually forms rounded clumps up to 18" high and 24" in diameter.

The leaves are needle-like, sessile, from ¼ to 1" long, grouped in 3's at the nodes and crowded near the ends of the branches. The margins are so revolute they almost cover the lower surface and appear as 2 parallel whitish lines.

The flowers are white or pink and ¼" or less across. They have 5 linear-shaped petals and grow in short axillary clusters. The fruits are spherical white drupes, about ¼" in diameter, containing 1 or 2 seeds.

The plants grow in brackish soil of open areas near salt water. They make an inspiring sight when in flower or in fruit.

Specimens were photographed in fruit on Sugarloaf and Big Pine Keys in July and August.

DESIGNATED STATUS: Endangered.

Suriana maritima L.   SURIANACEAE
Common Name: **Bay Cedar.**

This densely branched shrub or small tree is native to South Florida, the Keys, the Bahamas, the West Indies and Tropical America. Although it is usually seen 3 to 4' high, it does at times reach a height of 10 to 16'.

The wood is very hard and heavy. The bark is dark brown, rough and fissured. The paddle-shaped leaves, grayish or yellowish-green, are linear, up to 1½" long and ¼" wide. They are fleshy, somewhat pubescent, with entire margins, evergreen, alternate and clustered at the ends of the twigs.

The flowers are yellow, 5-petaled, solitary and about ½" wide. The fruit consists of 4 or 5 round, hairy, nut-like parts nestled in the persistent calyx, each containing 1 seed.

The plant is found throughout the Keys. The one pictured here was photographed on Geiger Key in October.

DESIGNATED STATUS: Endangered.

Swietenia mahagoni (L.) Jacq. MELIACEAE
Common Names: **Mahogany, West Indies Mahogany.**

One of the world's most valuable timber sources, this tree, growing to a height of 40 to 60', is native to South Florida, the Keys, the Bahamas, the West Indies and Tropical America. The wood is close-grained, heavy, hard, strong and a rich red-brown color. It is used for cabinets, ships and fine furniture. The bark has been used as a substitute for quinine.

The leaves are pinnately compound with 6 to 8 pairs of leaflets, briefly deciduous and 4 to 8" long. The leaflets are dark green above, yellowish or brownish underneath, leathery, ovate to ovate-lanceolate in shape, with entire margins and up to 4" long.

The flowers are greenish-yellow or white, about ⅛" long, 5-petaled in small clusters on stalks in the leaf axils. The fruit is a thick-walled, ovoid, woody, brown, erect pod, 4 to 5" long and 2½" in diameter. It splits open from the base upward releasing winged seeds.

This tree was photographed on Sugarloaf Key and in Key West. The flowers were taken in June.

DESIGNATED STATUS: Threatened.

Tecoma stans (L.) Juss.   BIGNONIACEAE
Common Name: **Yellow Elder.**

Native to Southeastern U.S.A., the West Indies and Tropical America, this shrub or small tree has become naturalized in the Keys and spreads weed-like in many areas. It grows erect up to 25' tall on side roads and hammocks throughout the Keys.

The leaves are pinnately compound with 5 to 13 light green leaflets. The leaflets are lanceolate to elliptic with serrated margins, 3 to 5" long and pointed at the apex. The flowers are bright yellow, bell-shaped, about 2" long, flared to about 1" across and appear in terminal clusters at the ends of the branches. The seed pod is a slender capsule up to 8" long that splits open when ripe and dry scattering the seeds. The opened pods remain on the trees for a period of time.

It was photographed in flower on Plantation Key and in Key West in January, May and October.

Thespesia populnea (L.) Soland. ex Correa   MALVACEAE
Common Names: **Portia Tree, Seaside Mahoe, Cork Tree, False Rosewood.**

This tree is a native of India but has become naturalized in South Florida and the Florida Keys. It reaches a height of 25' or more. The leaves are evergreen, heart-shaped or poplar-like, pointed and about 5" long. The flowers are hibiscus-like, about 3" across, cupped, light yellow with a red center in the morning and turning dark red in the late afternoon. They are edible raw as in a salad or dipped in a batter and fried.

    The fruit is a leathery, 5-sectioned, flattened capsule about 1½" across, turning black when mature and containing several 3-sided brown seeds. The juice of immature fruit and bark reportedly has been used medicinally, oil has been extracted from the seeds, and red dye has been made from the bark. The wood is colored in shades of red and purple.

    It was photographed on Sugarloaf Key in May in flower.

Thrinax morrisii H. Wendl.   ARECACEAE
Common Names: **Key Thatch Palm, Brittle Thatch Palm, Key Palm.**

This palm is now on the "Threatened List." It is native to the West Indies, the Bahamas, Central America and the Florida Keys. It has a smooth, gray, stout trunk and usually grows from 8 to 25' high. At times it has a swollen base composed of matted roots.

The palmate leaves are from 3½ to 4' across with many divisions cut about halfway to the base. They are shiny light-green above and silvery-white below. (The difference in color is more noticeable on this palm than on the *T. radiata*). The petioles are 3 to 4' long and spineless. The small white flowers on long arching inflorescences have very short pedicles. They are almost sessile. The fruit is a white single-seeded berry from ⅛ to ¼" in diameter and like the flower has almost no pedicle. In contrast, the flowers and fruit of the *T. radiata* have ¼" long pedicles and the fruit is larger.

In the past the leaves were used for thatching and the stems for fencing.

The plants were photographed on Big Pine Key and Sugarloaf Key in September.

SCIENTIFIC SYNONYMS: *T. keyensis* Sarg., *T. microcarpa* Sarg.
DESIGNATED STATUS: Threatened.

Thrinax radiata Lodd. ex J. A. & J. H. Schult.   ARECACEAE
Common Names: **Jamaica Thatch Palm, Florida Thatch Palm.**

This palm is native to South Florida, the Keys, the Bahamas, the West Indies and Central America. Its leaves were very useful to the Indians and the early settlers as thatch material for their homes. It is now on the "Threatened List." Reaching a height of 30' or more, it is usually characterized by a mass of roots enlarging the base.

The leaves are fan-shaped, almost circular, about 3' in diameter and cut halfway to the base into many segments. They are green above with prominent yellowish ribs and paler below. The petioles are 2 to 4' long and are unarmed.

The flowers are small, whitish in color, and grow in long clusters on branched stalks about 3' long. The fruits are almost globular, white in color, ¼" in diameter on a ¼" stem and containing a single smooth seed.

These pictures were taken on Sugarloaf Key in August.

SCIENTIFIC SYNONYMS: *T. parviflora* Sw., *T. floridana* Sarg.
DESIGNATED STATUS: Threatened.

Trema lamarckianum (Roem. & Schult.) Blume   ULMACEAE
Common Name: **West Indies Trema.**

This plant is native to South Florida, the Keys and the West Indies. It is very similar to the Florida Trema *(trema micranthum)* except the leaves are only about one-fourth as large and the flowers are whitish or pinkish-white. It grows as a shrub or small tree to 20'. It has thick, pubescent twigs.

The leaves are dull dark-green, flat, 1 to 2" long, alternate with sawtoothed margins. They are persistent, lanceolate, elliptic or ovate in shape with a very rough pubescent or velvety upper surface.

The flowers are both unisexual and bisexual on the same plant. They occur in axillary clusters and have 5 whitish or pinkish-white petals. The fruit is an ovoid drupe, pink in color, smooth and about ⅛" long.

The plant was photographed in Tavernier in May.

Trema micranthum (L.) Blume   ULMACEAE
Common Name: **Florida Trema.**

This evergreen sprawling shrub or small tree is native to South Florida, Tropical America, the West Indies and the Florida Keys. It has dark-brown bark and is usually shrubby and gangly with long horizontal branches, but occasionally develops a small erect trunk to 25 to 30'.

The leaves are simple, dull, velvety, dark-green above, lighter below, persistent and alternate on ½" petioles. They are ovate with an abruptly tapered apex and a heart-shaped or oblique-rounded base. They are 3 to 4" long, 1 to 2½" wide and have fine serrated or toothed margins.

The flowers are unisexual, both sexes appearing in the many-flowered, axillary clusters. They are greenish-yellow with 5 sepals and no petals. The fruits are yellow to orange, nearly spherical, fleshy, 1-seeded drupes less than ¼" in diameter and are a source of food for many birds.

These plants were photographed on Key Largo and Sugarloaf Key in February and April.

SCIENTIFIC SYNONYM: ***T. floridana*** Britton.

Vallesia antillana Woodson APOCYNACEAE
Common Names: **Pearl-berry, Tear-shrub.**

This shrub occasionally develops into a small tree with a definite trunk to 10 or 12' in height. It is native to South Florida and the Florida Keys. It is evergreen and has pale furrowed bark. The leaves are light green, dull, on thin shoots or twigs, alternate, elliptic to obovate-elliptic in shape and up to 3" long.

The flowers are produced in inflorescences along the branches throughout the year. They have 5 stamens and 5 white petals at the end of a ¼" corolla tube. Flowers and fruits frequently appear on the same plant at the same time. The fruits are white, 1-seeded, pear-shaped drupes with whitish flesh and ¼ to ½" long.

Some authorities list the *Vallesia glabra* as a separate species with a smaller corolla of shorter lobes and elliptic-lanceolate leaves.

This plant was photographed on Big Pine Key in October.

Ximenia americana L. OLACACEAE

Common Names: **Tallowwood, Hog Plum, Purge Nut.**

This sprawling and twisting shrub may reach a height of 15 to 20'. It is native to South Florida, the West Indies, the Bahamas and the Florida Keys. The branches are vine-like and have stout straight spines ½ to 1" long. The leaves are 1 to 3" long, yellowish-green above and below, alternate, oblong or elliptic, rounded or notched at the apex.

The flowers are yellowish-white, 4 petaled, ½" wide, fragrant, bell-shaped and usually occur in 3 or 4-flowered axillary clusters. The fruit is a bright yellow drupe, broad oval or nearly round, about 1" long, with a smooth skin, a large oval seed and fleshy pulp, edible raw or cooked. The roasted kernels are edible but have a purging effect if eaten in large quantities. The wood is hard, brown tinged with red, close-grained and sometimes used as a substitute for sandalwood.

The fruit was observed ripe on Sugarloaf Key from April to July.

Yucca aloifolia L.   AGAVACEAE

Common Names: **Spanish Bayonet, Aloe Yucca, Spanish Dagger.**

This plant is native to Southern U.S.A., Mexico, the West Indies and the Florida Keys. It has a thick erect stem, either singular or branching 20 to 25' high. The leaves are dense, stiff, bayonet-shaped, 12 to 24" long and 1 to 2" wide ending in hard needle-sharp tips. They are coarse in texture, dark green and persistent, remaining on the plant for many years either in the green or brown state.

    The fragrant flowers are creamy white, lightly tinged with purple, fleshy, drooping cup-like in erect terminal clusters 1 to 3' long. The fresh blooms are tender and edible raw in salads or dipped in batter and fried. It is pollinated by the Yucca moth whose larvae feed on the ripening seeds. The fruits are nearly cylindrical capsules, 2 to 3" long. They turn light brown to black and contain many small seeds.

    The plants occur throughout the Keys. The ones shown here are on Sugarloaf Key and were in flower from May through September.

Zanthoxylum fagara (L.) Sarg.   RUTACEAE
Common Names: **Wild Lime, Lime Prickly-ash.**

This evergreen tree is native to South Florida, the Keys, the Bahamas and the West Indies. It has sharp recurved spines either solitary or in pairs along the twigs. Its compound leaves are odd-pinnate with 7 to 15 leaflets ½ to 1½" long. The leaflets have marinal notches, winged rachis and are ovate to obovate in shape.

Although the plants are dioecious, the flowers are basically of the same design. The male has 4 stamens, but its 4 greenish-yellow petals are larger than those of the female. The flowers are borne on short axillary spikes and remain in the "bud" stage for a relatively long period of time before opening. The brown fruits are usually in pairs and open when ripe to expose small, single, shiny-black seeds that cling to the outer wall of the hull.

The picture of the plant in flower (bud stage) was taken on Key Largo in June.

SCIENTIFIC SYNONYM: *Fagara pterota* L.

Zanthoxylum flavum Vahl — RUTACEAE

Common Names: **Satinwood, Yellow-heart, Yellow-wood, West Indies Satinwood.**

This rare and endangered tree is native to Bermuda, the Bahamas, the West Indies and the Florida Keys. Specimens were photographed in the wild state by the author on Bahia Honda Key and Stock Island. It has been reported on the Marquesas Keys. The wood is hard, golden yellow, beautifully figured and very valuable for select cabinets and furniture. The National Champion is on Bahia Honda Key—3'7" in circumference, 20' in height, with a crown diameter of 30'.

The leaves are pinnately compound, light green, alternate and 6 to 9" long with 5 to 9 leaflets in opposite pairs. The leaflets are oblong to ovate with rounded tips, entire margins or minutely scalloped, 2 to 4" long and dotted with minute, translucent, glandular windows. Small male and female flowers appear in June and July (on separate trees) in terminal panicles, and have 5 greenish-white petals. The small, ovoid, brown fruits usually contain a single, shining black seed.

SCIENTIFIC SYNONYM: *Fagara flava* (Vahl) Krug & Urban.
DESIGNATED STATUS: Endangered.

SELECTED REFERENCES

1. Bell, C. Ritchie and Bryan J. Taylor. 1982. *Florida Wild Flowers.* Laurel Hill Press. Chapel Hill, NC.
2. Burkhalter, A. P., D. P. Tarver, J. A. Rodgers, M. J. Mahler and R. L. Lazor. 1978. *Aquatic and Wetland Plants of Florida.* Bureau of Aquatic Plant Research and Control. Florida Department of Natural Resources. Tallahassee, FL.
3. Bush, C. S. and J. F. Morton. 1976. *Native Trees and Plants for Florida Landscaping.* Bull. 193. Florida State Department of Agriculture. Tallahassee, FL.
4. Graf, Alfred Byrd. 1978. *Tropica.* Roehrs Company. East Rutherford, NJ.
5. Harrar, E. S. and J. G. Harrar. 1962. *Guide to Southern Trees.* Dover Publications Inc. New York, NY.
6. Little, E. L., Jr. 1978. *Atlas of United States Trees* Volume 5 *Florida.* U.S.D.A. Forest Service Miscellaneous Publication No. 1361, U.S. Government Printing Office. Washington, DC.
7. Little, E. L., Jr. 1976. *Rare Tropical Trees of South Florida.* Conservation Research Report No. 20. U.S.D.A. Forest Service. Washington, DC.
8. Little, E. L., Jr., Roy O. Woodbury and Frank H. Wadsworth. 1974. *Trees of Puerto Rico and the Virgin Islands.* Agriculture Handbook No. 449. U.S.D.A. Forest Service. U.S. Government Printing Office. Washington, DC.
9. Long, R. W. and O. Lakela. 1971. *A Flora of Tropical Florida.* University of Miami Press. Coral Gables, FL.
10. McCurrach, James C. 1960. *Palms of the World.* 1970 reprint by Horticultural Books, Inc. Stewart, FL.
11. Menninger, E. A. 1964. *Seaside Plants of the World.* Hearthside Press, Inc. New York, NY.
12. Menninger, E. A. 1962. *Flowering Trees of the World.* Hearthside Press, Inc. New York, NY.
13. Menninger, E. A. 1975. *Color in the Sky.* Horticultural Books, Inc. Stuart, FL.
14. Morton, J. F. 1974. *Wild Plants for Survival in South Florida.* Trend House. Tampa, FL.
15. Morton, J. F. 1971. *Plants Poisonous to People in Florida.* Hurricane House. Miami, FL.
16. Morton, J. F. 1974. *500 Plants of South Florida.* E. A. Seeman Publishing, Inc. Miami, FL.

17. Patterson, Jack and George Stevenson. 1977. *Native Trees of the Bahamas.*
18. Perkins, K. D. and W. W. Payne. 1978. *Guide to the Poisonous and Irritant Plants of Florida.* Florida Cooperative Extension Service. I. F. A. S., University of Florida and Florida State Museum. Gainesville, FL.
19. Sargent, C. S. 1961. *Manual of the Trees of North America.* Dover Publications, Inc. New York, NY.
20. Small, John Kunkel. 1933. *Manual of the Southeastern Flora.* University of North Carolina Press. Chapel Hill, NC.
21. Stevenson, George B. 1974. *Palms of South Florida.*
22. Stevenson, G. B. 1979. *Trees of the Everglades National Park and the Florida Keys.* Banyan Books, Inc. Miami, FL.
23. Sturrock, David and Edwin A. Menninger. 1946. *Shade and Ornamental Trees for South Florida and Cuba.*
24. Tomlinson, P. B. 1980. *The Biology of Trees Native to Tropical Florida.* Harvard Printing Office. Allston, MA.
25. Ward, D. B. 1978. *Rare and Endangered Biota of Florida,* Volume 5, *Plants.* University Presses of Florida, Gainesville, FL.
26. Watkins, J. V. 1970. *Florida Landscape Plants: Native and Exotic.* University of Florida Press. Gainesville, FL.
27. Wunderlin, Richard P. 1982. *Guide to the Vascular Plants of Central Florida.* University Presses of Florida. Gainesville, FL.

GLOSSARY

Achene A hard, dry indehiscent, one-seeded fruit with a single cavity.

Acuminate. Long-tapering; attenuated.

Acute Ending in a point which is less than a right angle, but not so tapering as "acuminate."

Aerial Epiphytic plants; plants or parts of plants living above the surface of the ground or water.

Alternate Said of leaves occurring one at a node, those of successive nodes forming a definite sequence around the stem; said also of members of adjacent whorls in the flower when any member of one whorl is in front of or behind the junction of two adjacent members of the succeeding whorl.

Anther The pollen-bearing part of a stamen.

Apex The tip or distal end.

Areole Small depression in the surface of cacti on which spines occur.

Aquatic Living in water.

Aril Fleshy material associated with or covering a seed.

Asexual Characterized by reproduction which does not involve the fusion of a sperm and an egg.

Axil The upper angle between an organ and the axis which bears it, such as the angle between the leaf and the stem bearing the leaf.

Axillary Growing in an axil.

Axis The main or central line of development of a plant, structure, or organ, such as the main stem.

Berry A fleshy fruit, few- to many-seeded.

Bipinnate Divided into pinnate segments with each segment again divided into pinnate segments.

Bisexual Having both sexes on the same individual plant; a hermaphrodite.

Bract A reduced or modified leaf, particularly the scale-like leaves in a flower cluster. Also said of any bractlike emergence.

Calyx The outermost whorl of the floral envelopes, composed of separate or united sepals; it may sometimes be petaloid.

Capsule A dry, dehiscent fruit originating from two or more carpels.

Carpel One of the foliar units of which a pistil is composed. If one carpel forms the pistil the latter is simple; if more than one, the pistil is compound.

Cauline Occuring along a stem, as opposed to basal.

Compound Formed of several parts united in one common whole, as is a compound pistil; or of leaves composed of two or more distinct leaflets.

Cone Seed bearing structure of the conifers.

Corolla The second whorl of the floral envelope, the units of which are petals; frequently the showy part of a flower.

Costapalmate Having a petiole or leafstem that continues on into the blade of the leaf and curves downward.

Crenate Margin with blunt-rounded teeth.

Crownshaft Green column of clasping leaf bases at the top of the palm, not found on all species.

Cuneate Wedge-shaped; tapering toward the point of attachment.

Deciduous Losing leaves seasonally.

Dehiscent Opening and shedding contents; said of fruits and stamens.

Dioecious Having staminate and pistillate flowers in different plants.

Distal Opposite the point of attachment; apical; away from the axis.

Drupe A fleshy, one-seeded indehiscent fruit containing a stone with a kernel; a stone-fruit such as a plum. Drupelet: a diminutive drupe. Drupaceous: resembling a drupe, possessing its character, or producing similar fruit.

Elliptic In the form of a flattened circle, usually more than twice as long as broad. Ellipsoid: an elliptic solid.

Endangered Species in imminent danger of extinction or exterpation and whose survival is unlikely if the causal factors presently at work continue operating.

Entire Having a margin devoid of any indentations, lobes or teeth; said of the margin of appendages such as bracts, stipules, sepals, petals and leaves.

Evergreen Never completely without leaves; new ones grow as old leaves drop.

Filament The stalk bearing the anther, or any thread-like structure.

Follicle A fruit, usually developing from a simple pistil and dehiscing along one margin.

Fruit The matured pistil or pistils and their accessory structures, bearing the ripened seeds.

Glabrous Without hairs or pubescence of any kind.

Glaucous Having a frosted or whitish waxy appearance from a waxy bloom or powdery coating.

Globose Shaped like a globe or sphere.

Glochid Tufts of very small spines.

Hammock Group of evergreen hardwood species in

an island surrounded by contrasting vegetation types.

Indehiscent Said of fruits that remain closed and do not shed their seeds.

Inflorescence An aggregation of flowers occurring clustered together in a particular manner which is usually characteristic of a given kind of plant.

Lanceolate Much longer than broad; from a broad base tapering to the apex; lance-shaped.

Leaflet A discrete segment of a compound leaf.

Linear Long and slender, with more or less parallel sides.

Lobe An outward projection from the margin of an organ, usually with the margin indented on either side of the projection, as in leaves.

Male (plant or flowers) Having stamens but no pistils.

Marginal Of, pertaining to, or attached to the edge.

Midrib The conspicuous central vein in the vascular system of an appendage.

Monoecious Having the stamens and pistils in different flowers on the same plant.

Native An original or indigenous inhabitant.

Naturalized Of foreign origin, but established and reproducing itself as though a native.

Oblanceolate Pointed at the apex, broadest above the middle, and tapering to the base.

Oblong Longer than broad, the sides nearly parallel for most of their length.

Obovate Ovate in shape, but with the broadest part near the distal end.

Obovoid Inversely ovoid.

Obtuse Having a blunt or rounded terminal part.

Opposite As said of leaves; occurring two at a node on opposite sides of the stem.

Ovary The part of the pistil bearing the ovules and maturing to form at least part of the fruit which bears the seeds.

Ovate Said of a plant structure having the shape of the outline of an egg.

Ovoid Egg-shaped.

Ovule An unfertilized egg.

Palmate Having several lobes radiating from a common base like the fingers from the palm of the hand; fan-shaped.

Panicle A compound inflorescene, that is, one in which the axis is branched one or more times.

Pedicel Stalk or stem of a flower in a flower cluster.

Peduncle The stem of a solitary flower or the main stem of a flower cluster.

Perfect Said of flowers that have both stamens and pistils.

Persistant Said of an organ that remains attached after ceasing to perform its usual biological function.

Petal A unit segment of the corolla.

Petiole The stem or stalk of a leaf.

Pinnate Having a common elongate rachis or axis, with segments arranged either oppositely or alternately along either side; feather-like.

Pinnately compound Said of structures the lateral segments of which are discrete and arranged along a common axis.

Pistil One of the essential organs of a flower, consisting usually of stigma, style and ovary, the ovary containing the ovule or ovules.

Pistillate Bearing pistils only.

Pneumatophore A specialized structure developed from the root in certain plants growing in swamps and marshes and serving as a respiratory organ.

Pollen The powdery grains which bear the sperm nuclei and which are contained in the anther.

Pubescent Hairy.

Raceme An inflorescence with a single axis, the flowers arranged along it on pedicles.

Radical Arising from the root or its crown; leaves that are basal or rosulate.

Rachis The prolongation of the peduncle through a flower cluster, or of a petiole through a compound leaf.

Rare Species with small populations in the state which, though not presently endangered or threatened as defined, are potentially at risk. Included are species that may be localized within a restricted geographical region or habitat or thinly scattered over a more extensive range.

Recurved Turned backward in a curving manner.

Resin An aromatic material secreted by plants.

Rosette A whorl of leaves radiating from a crown or center usually at or slightly above the ground.

Scorpoid A coiled inflorescence with flowers borne on one side alternately in opposing rows; coiled like a scorpion's tail.

Seed A mature ovule containing an embryo; in flowering plants developing within a fruit.

Sepal One of the segments of the calyx.

Serrate Having marginal teeth pointing forward.

Sessile Joined directly by the base without a stalk, pedicel or petiole.

Sheath The basal part of a lateral organ that closely surrounds or invests the stem.

Simple Neither branched nor otherwise compound.

Spatulate Shaped like a spatula.

Spike A type of inflorescence in which the axis is somewhat elongated and the flowers are numerous and sessile.

Spine A rigid, sharp-pointed structure usually modified from a stem.

Spinulose With small spines.

Stamen The pollen-bearing organ; usually consisting of the stalk or filament and the anther containing the pollen.

Staminate Said of plants or structures bearing stamens and not bearing pistils.

Stem The part of the plant bearing the foliar and floral organs.

Stigma That part of the pistil that receives the pollen and in which pollination is effected.

Stipule An appendage occurring at the leaf node.

Style A short or long, simple or branched stalk arising from the ovary and bearing the stigma or stigmas; the part of the pistil which connects ovary and stigma.

Succulent Fleshy; composed of soft, watery tissue.

Thorn A sharp-pointed modified branch.

Threatened Species believed likely to move into the endangered category in the near future if the causal factors now at work continue operating.

Toothed Bearing teeth.

Tree Woody plant having one erect stem or trunk at least three inches in diameter at a point $4\frac{1}{2}'$ above the ground, a more or less definitely formed crown of foliage and a height of at least 12 feet.

Tubular Cylindrical and hollow.

Umbel An inflorescence of few to many flowers on stalks of approximately equal length arising from the top of a scape of peduncle.

Unisexual In flowering plants, said of a plant or flower that either bears only stamens or only pistils, but not both.

Veins The ultimate branches or divisions of the vascular system, as in leaves or petals.

Whorl A ring of leaves, flower parts or flowers occurring at a single node.

IDENTIFICATION KEYS

The Identification Keys are based on characteristics of the leaves which can be observed in the Keys, generally, throughout the year. This feature, when linked with the identifiable characteristics of the flowers, fruit, spines, bark, etc. will result in positive identification.

The plants included in the book fall into nine major groups as follows:

I. Leaves feather-like or hand-shaped: PALM Family—Arecaceae.

II. Leaves or stems pine needle-like: PINE Family—Pinaceae and BEEFWOOD Family—Casuarinaceae.

III. Leaves very small and scale-like or absent, stems succulent, fleshy and spiny: CACTUS Family—Cactaceae.

IV. Leaves basal or cauline with stiff sharp tips, bayonet or dagger-like: AGAVE Family—Agavaceae.

V. Leaves—other—simple and alternate.

VI. Leaves—other—simple and opposite.

VII. Leaves—other—simple and clustered or whorled.

VIII. Leaves—other—compound and alternate.

IX. Leaves—other—compound and opposite.

(The Identification Keys on the following pages further subdivide and describe these groups.)

IDENTIFICATION KEYS
INDEX

The page numbers on the right refer to the following Identification Key which further subdivides and describes these groups.

		Page
I.	PALMS	
	A. Pinnate Leaves.	181
	B. Palmate Leaves.	181
	C. Costapalmate Leaves.	181
II.	PINE—CASUARINA	
	A. Leaves—needle shaped.	181
	B. Leaves—scale-like.	182
III.	CACTI	
	A. Stems—spiny pads.	182
	B. Stems—longitudinal ridges.	182
	C. Stems—3-angled.	182
IV.	AGAVE—YUCCA	
	A. Leaves—sword-like.	182
V.	LEAVES—SIMPLE—ALTERNATE	
	A. Flowers—yellow	
	1. Leaves 4" long or less.	183
	2. Leaves more than 4" long.	184
	B. Flowers—white	
	1. Leaves 4" long or less.	185
	2. Leaves more than 4" long.	187
	C. Flowers—red	
	1. Leaves 4" long or less.	189
	2. Leaves more than 4" long.	189
	D. Flowers—green	
	1. Leaves 4" long or less.	190
	2. Leaves more than 4" long.	190
VI.	LEAVES—SIMPLE—OPPOSITE	
	A. Flowers—yellow	
	1. Leaves 4" long or less.	191
	2. Leaves more than 4" long.	191
	B. Flowers—white	
	1. Leaves 4" long or less.	192
	2. Leaves more than 4" long.	194
	C. Flowers—red	
	1. Leaves 4" long or less.	195
	2. Leaves more than 4" long.	195
	D. Flowers—green	
	1. Leaves 4" long or less.	195

  E. Flowers—blue
   1. Leaves 4" long or less. 196
VII. LEAVES—SIMPLE—CLUSTERED OR WHORLED
  A. Flowers—white or pink
   1. Leaves 4" long or less. 196
  B. Flowers—greenish yellow—with purple streaks
   1. Leaves more than 4" long. 196
VIII. LEAVES—COMPOUND—ALTERNATE
  A. Flowers—yellow
   1. Leaves pinnate. 196
   2. Leaves bipinnate. 197
  B. Flowers—white
   1. Leaves pinnate. 198
   2. Leaves bipinnate. 199
  C. Flowers—red
   1. Leaves pinnate. 199
   2. Leaves bipinnate. 199
   3. Leaves trifoliate. 199
  D. Flowers—green
   1. Leaves pinnate. 200
   2. Leaves trifoliate. 200
IX. LEAVES—COMPOUND—OPPOSITE
  A. Flowers—yellow
   1. Leaves pinnate. 200
   2. Leaves bipinnate. 200
  B. Flowers—white
   1. Leaves pinnate. 201
  C. Flowers—blue
   1. Leaves pinnate. 201

(The Identification Keys on the following pages further subdivide and describe these groups.)

IDENTIFICATION KEYS

The numbers in parenthesis beside each plant species name indicate the page number in the text where further information is found.

I. PALMS
   A. Pinnate Leaves
      1. Leaflets in single horizontal plane, no crownshaft; green to yellow fruit (8" or more long)
Cocos nucifera (58)
      2. Leaflets in 2 rows forming "V" shaped leaves, short crownshaft; red fruit, (½ to 1" long)
Pseudophoenix sargentii (130)
      3. Leaflets in 4 different rows at different angles making leaves appear roundish, tall green crownshaft; red fruit (½ to 1" long)   *Roystonea elata* (142)

   B. Palmate Leaves
      1. Spines on petioles
         a. Single trunk, erect, inclined, underground, possibly branched; blue-black fruit (½ to ¾" long), may form large colonies   *Serenoa repens* (150)
         b. Multiple trunks, erect; black fruit (¼ to ½" long)
Accoelorrhaphe wrightii (6)
      2. No spines on petioles
         a. Leaves divided deeply, green above, silver-gray below; black fruit (approx. ½" long), trunk smooth and brownish-gray   *Coccothrinax argentata* (57)
         b. Leaves divided about halfway to base, light-green above, silvery-gray below; fruit white (⅛" or less long); very short pedicels—almost sessile
Thrinax morrisii (162)
         c. Leaves divided about halfway to base, green above, paler below; fruit white (approx. ¼" long) on ¼" pedicels   *Thrinax radiata* (163)

   C. Costapalmate Leaves
      1. No spines on petioles
         a. Leaves divided deeply, medium-green above, grayish-green below; trunk smooth, grayish and frequently covered with plaited old leafbases
Sabal palmetto (143)

II.   PINE—CASUARINA
   A. Leaves—Needle-shaped
      1. Leaves: dark-green, 8 to 12" long, clusters of 2 or 3
         Fruit· seed cones 2 to 6" long   *Pinus elliottii* (122)

    B. Leaves—Scale-like
      1. Leaves: very small, in whorls at nodes of needle-like jointed twigs, approx. 1/16" long
        a. Leaves: 6 to 8 in whorl
          Fruit: cone-like, woody, 1/2" long
                      Casuarina equisetifolia (41)
        b. Leaves: 10 to 16 in whorl
          Fruit: not known to occur in the Keys
                      C. glauca (42)

III. CACTI
    A. Stems—Flattened—Fleshy—Spiny Pads
      1. Flowers: yellow, 2 to 4" across
        Fruit: pear-shaped, red, purplish pulp
                      Opuntia stricta (120)
      2. Flowers: red, 1/2 to 1" across
        Fruit: obovoid, yellow   *O. spinosissima* (119)
    B. Stems—Cylindrical—Longitudinal Ridges & Grooves
      1. Stems: simple or branched, 9 or 10 ridges
        a. Stems: erect or reclining, young buds clothed with white hair
          Spines: 7 to 10 per areole   *Cereus gracilis* (44)
        b. Stems: upright columnar, young buds not clothed with white hair
          Spines: 4 to 25 per areole   *C. robinii* (46)
    C. Stems—3-Angled—Usually Reclining
      1. Stems: often supported by other plants, new growth rose-colored
        Spines: 4 to 7 per areole   *C. pentagonus* (45)

IV. AGAVE—YUCCA
    A. Leaves—Sword-like, terminating in hard, sharp spines, extending from rosette at base and along trunk
      1. Flowers and Fruit: on stalk up to 25' high
        a. Leaves: 5' or more long
          Marginal Spines: weak or absent
          Trunk: almost non-existent   *Agave sisalana* (8)
        b. Leaves: 3' or more long
          Marginal Spines: sharp and recurving
          Trunk: 2 to 8' high   *A. decipiens* (7)
      2. Leaves: 12 to 24" long
        Marginal Spines: spinulose-serrate
        Flowers: in terminal clusters 1 to 3' high
                      Yucca aloifolia (168)

V. LEAVES—SIMPLE—ALTERNATE
   A. Flowers—Yellow
      1. Leaves 4" long or less
         a. Leaves: heart-shaped, pubescent, deeply notched at base, scalloped margins
            Flowers: 5-petaled, 1½" across
            Abutilon permolle (1)
         b. Leaves: rough veins on upper surface, margins entire or faintly serrate
            Flowers: small, greenish-yellow on spikes 1 to 2" long, fragrant
            Fruit: dark-brown, ⅓" diameter capsules
            Ateramnus lucidus (15)
         c. Leaves: ovate or spatulate 1 to 2" long
            Flowers: pale yellow, ¼" wide
            Fruit: 3-lobed capsule, ¼" wide
            Argythamnia blodgettii (14)
         d. Flowers: greenish-yellow, 5-petals
            Fruits: 3-lobed drupe
            Leaves: ovate to elliptic
            (1) Leaves: strongly 3-nerved at base
                Margins: serrate   *Colubrina asiatica* (60)
            (2) Leaves: pinnately nerved, margins entire
                (a) Leaves: thick, black glands on under surface, young stems pubescent with rusty hairs
                    C. arborescens (59)
                (b) Leaves: thin, stems sparsely pubescent, small glands on lower part of leaf margins
                    C. elliptica (61)
         e. Leaves: oval, light green, minutely toothed along margins, strong yellow midrib
            Flowers: greenish-yellow on 6" spikes
            Fruit: greenish-yellow, crabapple shape
            Hippomane mancinella (97)
         f. Leaves: oblong or elliptic, notched at apex
            Flowers: ¼" wide, axillary hairy-stemmed drooping clusters
            Fruit: brown, scurfy skin, round 1½" diameter, spongy flesh, copious latex
            Manilkara bahamensis (110)
         g. Leaves: elliptic-lanceolate, irregularly toothed, 2 to 3" long, aromatic
            Flowers: small in axillary clusters
            Fruit: round, ⅛" diameter, covered with bluish wax
            Myrica cerifera (116)

h. Leaves: fleshy, paddle-shaped, entire, clustered at end of twigs, 1 to 1½" long
Flowers: 5 petals, flared, ½" across, solitary or small clusters   *Suriana maritima* (158)
i. Leaves: oblong or elliptic, apex rounded or notched, 1 to 3" long
Flowers: 4 petals, ½" wide, fragrant, bell-shaped
Fruit: yellow drupe, nearly round, 1" diameter
Ximenia americana (167)
j. Leaves: oblanceolate to spatulate, ½ to 1½" long, tip rounded, base wedge-shaped
Flowers: ⅛" long, fragrant
Fruit: ¼ to ½" long, cylindrical blue-black berry
Bumelia celastrina (24)
k. Leaves: ovate or elliptic, 1 to 3" long, margins rolled under
Flowers: 5 petals, 1" across
Fruit: 3 to 5 valve capsule
Cienfuegosia yucataniensis (51)
2. Leaves more than 4" long
  a. Leaves: narrowly elliptic or spatulate, margins entire or faintly notched
Flowers: no petals, greenish yellow, ⅛" across
Fruit: yellowish 3-celled capsule with wings, turns red and brown   *Dodonea viscosa* (72)
  b. Leaves: variable, usually oblong or oval, 3 to 5" long, margin entire or slightly notched
Flowers: small in dense clusters
Fruit: ovoid drupe, white, ½ to 1" long
Drypetes diversifolia (73)
  c. Leaves: heart-shaped, nearly round, short tip, 4 to 6" wide
Flowers: yellow flowers turn red in afternoon
Fruit: long pointed capsule, ¾" wide, brown
Hibiscus tiliaceus (96)
  d. Leaves: yellow-green, margins wavy, elliptic to ovate, 2 to 8" long, unpleasant odor
Flowers: light-yellow, clustered in leaf axils
Fruit: yellow, globular, 1" long
Mastichodendron foetidissimum (112)
  e. Leaves: oblong, elliptical or obovate, margins entire, 2 to 5" long
Flowers: light-yellow on 2 to 3" catkins
Fruit: acorns, 1" long, enclosed ¼ of length in brown cup ⅓" in diameter   *Quercus virginiana* (136)

 f. Leaves: heart-shaped, poplar-like, 5″ long, sharp-pointed
  Flowers: hibiscus-like, yellow in a.m. turning dark-red in p.m., 3″ across
  Fruit: 5 sectioned capsule, 1½″ across, flattened
        Thespesia populnea (161)
 g. Leaves: lanceolate, obtuse at base, 2 to 5″ long, fine serrated margins, pubescent above
  Flowers: small, greenish-yellow, on axillary stalks
  Fruit: reddish-orange, globular, ⅛″ diameter
        Trema micranthum (165)
 h. Leaves: palmate, 7 to 9 lobes, to 24″ wide
  Flowers: 5 petals, greenish-yellow, 1″ across
  Fruit: melon-like, yellow, 3″ or more long
        Carica papaya (37)
B. Flowers—White
 1. Leaves 4″ long or less
  a. Flowers: heads with many florets, not fragrant
   Fruit: in clusters of ribbed achenes with attached hairs
   (1) Leaves: 1 to 3½″ long, narrowly linear, less than ⅕″ wide, entire or toothed
        Baccharis angustifolia (17)
   (2) Leaves: 1 to 3″ long, lanceolate-ovate to spatulate, more than ⅕″ wide, deeply toothed or notched   *B. halimifolia* (18)
  b. Flowers: corolla deeply cut, 5-lobed
   Fruit: orange drupe
   (1) Leaves: 2 to 3″ long, broadly ovate, to 2″ wide, some notched at tip *Bourreria ovata* (22)
   (2) Leaves: narrowly ovate to obovate, less than 1½″ wide, tips rounded, obtuse, infrequently notched
    (a) Fruit ¼″ diameter, leaves ½ to ¾″ long, short petioles   *B. cassinifolia* (21)
    (b) Fruit ⅖″ diameter, leaves 1 to 3″ long, ¼″ petiole   *B. radula* (23)
  c. Leaves: elliptic-ovate, 2 to 3″ long, rounded at tip, crenate margins, petioles often winged
   Flowers: axillary clusters
   Fruit: greenish-yellow, globular, 1 to 2″ diameter
        Citrus aurantifolia (53)

d. Leaves: elliptic to lanceolate-ovate, saw-toothed margins, softly hairy, ¾ to 1½" long
Flowers: in scorpioid cymes, corolla funnel-shaped, ¼" long
Fruit: ovoid drupe, red, ¼" diameter
Cordia globosa (64)

e. Leaves: narrowly-linear, smooth above, hairy below, 1½ to 3" long
Flowers: on 2 to 4" racemes
Fruit: subglobose capsule, yellowish, ¼" long
Croton linearis (69)

f. Leaves: lanceolate to ovate, margins entire, veins prominent, 3 to 4" long
Flowers: small in axillary clusters
Fruit: red, globular, downy, ¼" diameter
Drypetes lateriflora (74)

g. Leaves: linear to spatulate, pale silky pubescence, 2 to 4" long
Flowers: in dense, one-sided cymes with 2 to 4 recruved branches
Fruit: very small, black, ovoid drupe
Mallotonia gnaphalodes (109)

h. Leaves: thick, oblong or elliptical, apex rounded or minutely notched, margin entire or wavy
Flowers: greenish-white, small in axillary clusters, 1 to 1½" long
Fruit: egg-shaped, bright red, ¼" long
Maytenus phyllanthoides (113)

i. Leaves: obovate, pointed, prominent veins, 2 to 4" long
Flowers: 5 or 6 petals, star-shaped, ¼" across, leaf axils
Fruit: green to yellow, spherical, bumpy surface, 1" diameter
Morinda royoc (115)

j. Leaves: thick, fleshy, clustered at end of twigs, obovate or spatulate, entire margins, 1 to 2½" long
Flowers: 5 or 6 lobes, shaped like a hand
Fruit: round shiny black drupe, ¼ to ¾" diameter
Scaevola plumieri (146)

k. Leaves: lanceolate, elliptic or ovate, veins prominent, margins serrate, downy, 1 to 2" long
Flowers: very small in dense axillary clusters
Fruit: pink, ovoid drupe, ⅛" diameter
Trema lamarckianum (164)

l. Leaves: elliptic to obovate-elliptic, 1 to 3″ long
Flowers: 5 petals, ½″ wide, in axillary clusters
Fruit: white, pear-shaped drupes, ¼ to ½″ long
Vallesia antillana (166)

m. Leaves: oval, nearly round, slightly notched at apex, 1 to 3½″ long
Flowers: 5 petals in short axillary clusters
Fruit: globular, white or purple, 1 to 1¾″ long
Chrysobalanus icaco (49)

n. Leaves: linear-lanceolate, usually in clusters, succulent, margins entire
Flowers: white to pale-lilac, 5 petals, axillary clusters ¼″ long
Fruit: red berry, egg-shaped, up to ½″ across
Lycium carolinianum (107)

o. Leaves: clasping stem, margins entire, ovate to ovate-lanceolate, longer than broad, 2 to 5″ long
Fruit: dark purple to black, round ¼ to ½″ long drooping clusters   *Coccoloba diversifolia* (55)

p. Leaves: leathery, elliptic to oblong-obovate, 2 to 4″ long, entire or lightly serrate margins above midpoint
Flowers: dioecious, 4 each petals, sepals, and stamens, about ¼″ across
Fruit: red or orange-yellow, ¼ to ½″ diameter
Ilex cassine (99)

2. Leaves more than 4″ long

a. Leaves: oval, short-pointed or rounded, margins entire, 6 to 8″ long
Flowers: creamy-white to purple, bell-shaped, 2½″ long
Fruit: oval to 4″ long   *Amphitecna latifolia* (10)

b. Leaves: oblong to elliptical, entire, 3 to 5″ long
Flowers: yellowish-white with red blotch in throat, trumpet-shaped, 1″ across
Fruit: yellow-green with light-brown blotches, inverted pear-shape, 2 to 5″ long
Annona glabra (12)

c. Leaves: lance-shaped, tip blunt-pointed, base wedge-shaped, 4 to 6″ long
Flowers: branched terminal stalks, fragrant, ¼″ wide
Fruit: black drupes, ¼″ across
Ardisia escallonioides (13)

d. Leaves: lanceolate, curving downward, tip pointed, 3 to 5" long
Flowers: 5-petaled, fragrant, short-stalked, along branches, ¼" across
Fruit: black, globular, ¼" long
Bumelia salicifolia (25)

e. Leaves: oval or linear to broadly obovate, tip rounded or notched, margins entire, 2 to 6" long
Flowers: 4 petals and many long stamens, fragrant, greenish-yellow seed pods 6 to 12" long
(1) upper stems and lower surface of leaves densely scaley, brown seeds in red pulp
Capparis cynophallophora (35)
(2) upper stems and lower surface of leaves glabrous, white seeds in red pulp   *C. flexuosa* (36)

f. Leaves: elliptic with pointed tip, clustered at twig ends, 2 to 5" long
Flowers: leaf axils, ½" across
Fruit: rough skin, sweet pulp, almost round, 2 to 3" diameter   *Manilkara zapota* (111)

g. Leaves: oblong to elliptical, blunt-pointed, entire recurved margins, 2 to 5" long
Flowers: borne on short spurs scattered along stem
Fruit: round black ¼" drupe
Myrsine floridana (117)

h. Leaves: elliptical or narrowly-oval, margins entire, aromatic, 3 to 6" long
Flowers: fragrant, ¼" across, in many-flowered inflorescences
Fruit: nearly round, blue-black in red cups, ½" long
Nectandra coriacea (118)

i. Leaves: lanceolate to ovate-elliptic tapering to a point, margins wavy
Flowers: ⅛" diameter in 1 to 3" long racemes
Fruit: spherical, somewhat flattened, red berry, ⅛" diameter   *Rivina humilis* (141)

j. Leaves: oval or elliptic, pubescent, wavy margins
Fruit: globular ¼ to ½" diameter
(1) Fruit: red on erect stems
Solanum donianum (153)
(2) Fruit: yellow   *S. erianthum* (154)

k. Leaves: clasping stem, smooth, margins entire
  (1) Leaves: ovate to ovate-lanceolate, longer than broad, 2 to 5" long
    Fruit: dark purple to black, round, ¼ to ½" long, drooping clusters *Coccoloba diversifolia* (55)
  (2) Leaves: suborbicular, broader than long, 4 to 8" long, wavy margins, new growth red
    Fruit: red to purple, ¾" long in grape-like clusters *C. uvifera* (56)

C. Flowers — Red (Pink-Lavender)
  1. Leaves 4" long or less
    a. Leaves: deltoid-ovate, irregular-toothed, 3-lobed margins, underside pubescent, 1 to 3" long
      Flowers: 1" long, cylindrical, nodding
      Fruit: ¼ to ½" capsule, pubescent
      Hibiscus poeppigii (95)
    b. Leaves: lanceolate-ovate to elliptic-ovate, pubescent, margins serrate, short petioles, almost sessile, 1½ to 2½" long
      Flowers: small, pink *Pluchea odorata* (128)
    c. Leaves: ovate-elliptic, margins entire, wavy, 1 to 2½" long
      Flowers: fragrant, 4 or 5 petals, ⅛" across, axillary or terminal clusters
      Fruit: ovoid, ¼ to ½" diameter, red turning black
      Schoepfia chrysophylloides (149)
  2. Leaves more than 4" long
    a. Leaves: obovate, rounded or slightly notched at tip, 3½ to 5" long
      Flowers: red-white or purplish-rose, ⅛" diameter
      Fruit: red berry, ⅛" diameter
      Canella winterana (34)
    b. Leaves: broad-oval, rough, pointed tip, 5 to 10" long
      Flowers: red-orange, 1 to 1½" across, flared trumpet-shape, flat terminal clusters
      Fruit: white, pear-shaped, fragrant, 1 to 1¼" long
      Cordia sebestena (65)
    c. Leaves: lanceolate-ovate to elliptic-ovate, pubescent, 3 to 6" long, definite petioles ¼ to ½" long, margins entire or obscurely serrate
      Flowers: pink *Pluchea symphytifolia* (129)

    d. Leaves: oblong or elliptical, 2 to 5" long, midrib prominent and yellow
       Fruit: red, berry-like, ½" across
       (1) Leaves: margins not recurved
          Fruit: sessile, leaf axils   *Ficus aurea* (85)
       (2) Leaves: margins usually recurved
          Fruit: ¼" stems   *F. citrifolia* (86)
    e. Leaves: lanceolate to oblong, wavy margins, rough textured
       Flowers: 5 petals, lavender, star-like
       Fruit: red, globular, ¼" long, drooping
                      Solanum bahamense (152)

D. Flowers—Green
  1. Leaves 4" long or less
    a. Leaves: ovate, obovate or elliptic, margins entire, 1 to 4" long
       Flowers: small, greenish, no petals, dense cone-like heads
       Fruit: purplish-green cones
       (1) Leaves: yellow-green, smooth
                    Conocarpus erectus (62)
       (2) Leaves: silvery-gray pubescent
                  C. erectus var *sericeus* (63)
    b. Leaves: ovate or obovate, tip rounded or blunt-pointed, 1 to 2" long, margins entire
       Flowers: 5 greenish petals
       Fruit: 3-lobed brown capsule, ¼" long
                    Savia bahamensis (145)
    c. Leaves: elliptical or oval, rolled under margins, 2 to 3" long, tip pointed or round
       Flowers: 4 petals, ¼" wide, axillary clusters
       Fruit: red drupes, ¼" diameter
                Schaefferia frutescens (147)
  2. Leaves more than 4" long
    a. Leaves: oval, tip short-pointed or rounded, green above, golden brown and downy below
       Flowers: ⅛" long, fragrant, short stalked
       Fruit: dark-purple, oval berry, ¾" long
                Chrysophyllum oliviforme (50)
    b. Leaves: elliptic to oblong-lanceolate, margins entire, aromatic when crushed
       Flowers: small, pale green or yellow, no petals
       Fruit: blue-black, ovoid, ⅓" long
                      Persea borbonea (121)

VI. LEAVES—SIMPLE—OPPOSITE
  A. Flowers—Yellow
    1. Leaves 4" long or less
      a. Leaves: oblanceolate or spatulate, pubescent, small pointed tip
        Flowers: composite heads, ½ to 1" across
        (1) Phyllaries spine tipped, reflexed
            Borrichia frutescens (20)
        (2) Phyllaries not spine-tipped, oppressed
            B. arborescens (19)
      b. Leaves: elliptical or oval, sharp-pointed
        Flowers: bell-shaped, ¼" long, fragrant
        Fruit: white globular drupes, ¼" diameter
            Chiococca alba (47)
      c. Leaves: elliptic or spatulate, blunt or pointed tips, margins entire
        Flowers: no petals, clusters along stems
        Fruit: black oval drupe, ¼ to ½" long
            Forestiera segregata (87)
      d. Leaves: oblong to obovate, obscure veins, thickened wavy margins, tip rounded
        Flowers: small, yellow-green, purple tinge
        Fruit: red, oval, ¼ to ½" diameter
            Guapira discolor (90)
      e. Leaves: paddle-shaped, obscure veins, curled under margins, rounded or notched tip
        Flowers: pale-yellow, fragrant, terminal clusters
        Fruit: globular berry, ⅓" diameter, orange-red
            Jacquinia keyensis (100)
      f. Leaves: oblong-oval or obovate, entire margins notched at tip
        Flowers: greenish-yellow, no petals, axillary, ¼" wide
        Fruit: oval or round, dark-purple, ½" long
            Reynosia septentrionalis (138)
    2. Leaves more than 4" long
      a. Leaves: 3-lobed, pubescent, long petioles
        Flowers: 5 petals, red spots at base, 2 to 3" wide
        Fruit: rough, triangular-shaped capsule, seeds with long wooly fibers   *Gossypium hirsutum* (88)
      b. Leaves: ovate to elliptic, tip blunt-pointed or rounded
        Flowers: 4 petals, 1" across, fragrant
        Fruit: brown, 1" long, conical, developing seedling 6 to 12" long while still on tree
            Rhizophora mangle (139)

B. Flowers—White
1. Leaves 4" long or less
   a. Leaves: obovate, tips rounded or small point
      Flowers: 5 petals, white turning pink to red
      Fruit: globular, brown, ¼" diameter
      Byrsonima lucida (27)
   b. Leaves: elliptic-ovate, long tapering point, margin entire, aromatic
      Flowers: no petals, many stamens
      Fruit: purplish-black berry, ⅓" diameter
      (1) Leaves: midrib depressed, new growth faintly downy   *Calyptranthes pallens* (32)
      (2) Leaves: midrib slightly raised, new growth shiny—not downy   *C. zuzygium* (33)
   c. Leaves: ½" long, obovate, margin entire
      Spines: 2 sharp spines, ¼ to 1" long in leaf axils
      Flowers: funnel-shaped, 4 flaring petals
      Fruit: small, globose white berry
      Catesbaea parviflora (43)
   d. Leaves: elliptic or ovate, 1 to 2" long
      Flowers: 5-lobed, axillary racemes, ¼" wide
      Fruit: white globular drupe, ¼ to ½" diameter
      Chiococca parvifolia (48)
   e. Leaves: oval, blunt-pointed, obscure veins
      Flowers: 5 petals, star-like, ¼" across
      Fruit: globular, ¼" diameter, purple-black
      Erithalis fruticosa (76)
   f. Flowers: axillary clusters
      (1) Pedicels of flowers and fruit short and stout—less than ⅕"
         (a) Leaves: rounded or obtuse at apex
            Eugenia foetida (81)
         (b) Leaves: acute or pointed at apex
            (11) Pedicels: stout, shorter than flowers in dense clusters   *E. axillaris* (79)
            (22) Pedicels: slender, longer than flowers, flowers in few-flower clusters
               E. rhombea (82)
      (2) Pedicels of flowers and fruits long and slender—over ⅕"
         (a) Leaves: narrowed at apex to prolonged tip
            Fruit: ⅕ to ⅓" diameter
            Eugenia confusa (80)

(b) Leaves: acute at apex, not narrowed to prolonged tip
Fruit: ⅓ to ⅔" diameter, yellow turning red, then black *E. rhombea* (82)

g. Leaves: oblong-lanceolate, sharp-pointed tip, conspicuous mid-veins
Flowers: white or pinkish turning orange, 3" flaring tube, 5 lobes, fragrant
Fruit: woody capsule, black when dry
Exostema caribaeum (83)

h. Leaves: elliptic to ovate or obovate, tip short pointed or blunt, prominent veins, pubescent below
Flowers: white, pink or reddish, 4 to 5-lobed corolla, ¼" long
Fruit: globular, flattened at tip, dark red, ⅖" across, downy *Guettarda elliptica* (91)

i. Leaves: oblong-ovate or elliptic, entire or slightly crenate margins, tip rounded or minutely notched, on square-stemmed twigs
Flowers: 5 spreading petals, long stalks
Fruit: blue-black, ovoid, ¼" long
Gyminda latifolia (93)

j. Leaves: oval, rounded at base and tip
Flowers: small, 5 greenish-white petals
Fruit: obovoid, 10-ribbed, ½" long
Laguncularia racemosa (102)

k. Leaves: narrowly elliptic to ovate, toothed margins, pubescent
Flowers: involucrate heads, 1" across
Fruit: globular, purple, ⅛" diameter
Lantana involucrata (104)

l. Leaves: ovate, reddish veins underneath
Flowers: white or pink, ½" wide, fragrant
Fruit: globose, purple-black, ¼ to ½" diameter on long stalks *Psidium longipes* (132)

m. Leaves: elliptic to ovate, rounded apex
Flowers: small, terminal or axillary clusters
Fruit: globose, red, ¼" diameter
Psychotria punctata (135)

n. Leaves: elliptic to nearly round, blunt pointed with small sharp point
Flowers: 5 petals, clustered in leaf axils or along stems, fragrant, ½" across
Fruit: round, white, ⅓" long with blue-black pulp
Randia aculeata (137)

2. Leaves more than 4" long
   a. Leaves: elliptic, green above, gray and salty below
      Roots: pneumatophores
      Flowers: 4-lobed, fragrant, ½" across
      Fruit: green, flattened egg-shaped, 1½"
                            Avicennia germinans (16)
   b. Leaves: obovate to cuneate, midrib and veins prominent, margins curved under
      Flowers: tubular, star-like flared petals, solitary or in several-flowered axillary terminal clusters
      Fruit: greenish-yellow, black spots, egg-shaped, 2 to 3" long   *Casasia clusiifolia* (38)
   c. Leaves: oblong-obovate, margins entire, often wavy
      Flowers: tubular, 5-lobed, axillary racemes
      Fruit: globular, red-brown, ¼ to ½" diameter
                            Citharexylum fruticosum (52)
   d. Leaves: obovate, rounded and notched at apex, margin entire, stiff
      Flowers: white petals, tinted pink, to 3" wide
      Fruit: globular, to 3" diameter, opens claw-like
                                     Clusia rosea (54)
   e. Leaves: elliptic, ovate or obovate, shiny and rough above, downy below
      Flowers: tubular, fragrant ½" long
      Fruit: globular, red, downy, ¼" diameter
                               Guettarda scabra (92)
   f. Leaves: elliptic or oblong, conspicuous veins, impressed above, raised below
      Flowers: 4 or 5 lobes, 1 to 1½" across
      Fruit: spherical, yellow, 1½ to 2" diameter
                               Psidium guajava (131)
   g. Flowers: small, terminal or axillary clusters, corolla funnelform, 4 to 5 lobed
      Fruit: globose or oblong drupe, red, ¼" diameter
     (1) Leaves: elliptic to oblong, pointed
         Inflorescence: sessile in leaf axils
                               Psychotria nervosa (134)
     (2) Leaves: lanceolate to oblanceolate
         Inflorescence: distinct stalks in upper leaf axils
                           Psychotria ligustrifolia (133)

C. Flowers—Red
1. Leaves 4" long or less
   a. Leaves: elliptical to ovate
      Flowers: very small, greenish to red
      Fruit: red globular drupe, ¼" diameter
      (1) Leaves: ½" long, margins toothed
      Crossopetalum ilicifolium (67)
      (2) Leaves: ½ to 1½" long, margins crenate or entire
      C. rhacoma (68)
   b. Leaves: 1" long or less, narrowly-elliptic, frequently clustered
      Flowers: tubular, 4 to 6 lobes, pink or yellowish, curled back
      Fruit: round yellow drupe, ¼" long
      Ernodea littoralis (77)
   c. Leaves: obovate, tips rounded or small point
      Flowers: 5 petals, white turning pink to red
      Fruit: globular, brown, ¼" diameter
      Byrsonima lucida (27)
2. Leaves more than 4" long
   a. Leaves: pointed at both ends, rough texture, finely toothed margins
      Flowers: pale pink, rarely white, 4 petals, ¼" long, leaf axil clusters
      Fruit: pink, violet, rarely white, ⅛" globular drupe
      Callicarpa americana (31)
   b. Leaves: elliptic, pointed at apex and base
      Flowers: tubular, ¼ to ½" long, 5 lobed, tassel-like clusters 5" across   *Hamelia patens* (94)
   c. Leaves: rough, ovate, apex pointed, toothed margins, aromatic
      Stem: hairy, prickly
      Flowers: white, yellow, pink turning orange and red, flat clusters 1 to 1½" wide
      Fruit: globular drupe, dark blue, ¼" diameter
      Lantana camara (103)
D. Flowers—Green
1. Leaves 4" long or less
   a. Leaves: broadly oval, blunt pointed, may be slightly notched, margins wavy
      Flowers: no petals, ¼" across, axillary clusters, yellowish-green
      Fruit: black spherical drupe, ⅓" diameter
      Krugiodendron ferreum (101)

  b. Leaves: elliptic or ovate, margins entire
   Spines: sharp, paired, recurved, axillary
   Flowers: small, no petals, axillary clusters
            Pisonia aculeata (124)
  c. Leaves: obovate or elliptic-ovate, margins entire, blunt-pointed, depressed veins
   Flowers: small, no petals, dense clusters
   Fruit: obovoid, less than ¼″ diameter
            P. rotundata (125)
E. Flowers—Blue
 1. Leaves 4″ long or less
  a. Leaves: ovate to elliptic, tip blunt or short pointed, margins serrate or entire
   Flowers: blue, purple or white, ½″ across
   Fruit: orange round drupes, ⅓″ diameter
            Duranta repens (75)
  b. Leaves: oblong or ovate-lanceolate, toothed margins
   Flowers: ⅜″ long, quill-like spikes
          Stachytarpheta jamaicensis (156)

VII. LEAVES—SIMPLE—CLUSTERED OR WHORLED
 A. Flowers—White or Pink
  1. Leaves 4″ long or less
   a. Leaves: needle-like, ¼ to 1″ long, grouped in 3's at nodes and ends of branches, sessile revolute margins
    Flowers: 5 petals, ¼″ wide, axillary clusters
    Fruit: spherical, white drupes, ¼″ diameter
           Strumpfia maritima (157)
 B. Flowers—Greenish Yellow—With Purple Streaks
  1. Leaves more than 4″ long
   a. Leaves: spatulate to oblanceolate, 2 to 4 clustered on short spurs
    Flowers: 5 petals, 2½″ wide, along branches
    Fruit: subglobose, 4 to 8″ diameter, hard-shelled, green turning brown *Crescentia cujete* (66)

VIII. LEAVES—COMPOUND—ALTERNATE
 A. Flowers—Yellow
  1. Leaves Pinnate
   a. Leaflets: 3 to 5 pairs, oblong-lanceolate or oval, ½ to 2¼″ long
    Flowers: butterfly-shape, 1″ across, axillary clusters
    Fruit: flat brown pods, 1 to 3″ long
           Cassia chapmanii (39)

b. Leaflets: up to 8 pairs, obovate-pointed, ¼ to ½" long
   Flowers: 5 flaring petals, orange markings near base, axillary
   Fruit: pea-shaped pod, 2" long
   Cassia keyensis (40)
c. Leaflets: 10 to 14, oval, oblong or obovate, round tipped or notched, 2 to 4" long
   Flowers: 4 to 5 petals, ¼" wide, terminal or axillary clusters
   Fruit: oval, 1" drupe, red to dark purple
   Simarouba glauca (151)
d. Leaflets: 11 to 25, elliptic or oval, recurved, ½ to 2½" long, margins entire
   Flowers: pea-shaped, 1" wide, terminal spikes 4 to 16" long
   Fruit: seed pod 2 to 8" long, ½" diameter
   Sophora tomentosa (155)
e. Leaflets: 6 to 8 pairs, ovate-lanceolate, 1 to 4" long, margins entire
   Flowers: 5 petals, ⅛" long, axillary clusters
   Fruit: brown, ovoid, woody, erect pod, 4 to 5" long, 2½" diameter   *Swietenia mahagoni* (159)
f. Leaflets: up to 10 opposite pairs, ovate, about 2½" long
   Flowers: fluffy globular clusters
   Fruit: flat tan pods, 8" long, loose seeds
   Albizia lebbeck (9)

2. Leaves Bipinnate
   a. Leaves: feathery on slender branches
      Flowers: globular ball-shaped heads, ¼ to 1" diameter
      Fruit: smooth dark-brown legumes, 1¼ to 3" long
      (1) Pinnae: 2 to 6 pairs
          Leaflets: ¼" long
          Legumes: cylindrical, rounded ends
          Spines: 1 to 1½" long   *Acacia farnesiana* (3)
      (2) Pinnae: 3 to 6 pairs
          Leaflets: ⅛" long
          Legumes: cylindrical, beaked ends
          Spines: ⅔ to 1" long   *A. pinetorum* (5)
      (3) Pinnae: 6 to 18 pairs
          Leaflets: ⅛" long
          Legumes: laterally compressed
          Spines: 1½" long   *A. macracantha* (4)

(4) Pinnae: 1 to 3 pairs
Leaflets: ⅓ to 1⅓" long
Legumes: pointed end, garden-pea shape
Spines: no spines (possibly spiny stipules)
A. choriophylla (2)

B. FLOWERS—WHITE
 1. Leaves Pinnate
   a. Leaflets: 6 to 12, oblong, round-tipped, 2½ to 8" long, low rounded-toothed margins
   Flowers: ¼" wide, fragrant, branching inflorescence, axillary
   Fruit: 3-lobed brown capsule, ⅘" long
   Cupania glabra (70)
   b. Leaflets: 1 to 4" long, ovate to elliptic
   Flowers: white or pinkish in axillary clusters, corolla ¼" wide
   Dalbergia brownii (71)
   c. Leaflets: 2 to 4, oblong or oval, 4 to 5" long, rounded or minutely notched at tip
   Flowers: 5 petals, fragrant, ¼" wide
   Fruit: globular, ½" long, red to purple
   Exothea paniculata (84)
   d. Leaflets: blades alternate on stem, narrowly lanceolate, grass-like, 2 to 8" long
   Stem: glabrous, main stem or culm woody, viny, up to 10 to 12' long
   Lasiacis divaricata (105)
   e. Leaflets: 3 to 7, usually 5, ovate, tip rounded or blunt pointed, 3 to 4" long
   Flowers: creamy-white in axillary panicles
   Fruit: oval, ½" long, dull orange
   Metopium toxiferum (114)
   f. Leaflets: 4 to 9 elliptical to lanceolate, pointed tips, 2 to 4" long, winged rachis
   Flowers: 4 to 5 petals, terminal clusters
   Fruit: globular orange berry, ¼ to ½" long
   Sapindus saponaria (144)
   g. Leaflets: 3 to 11 lanceolate to elliptic, 1 to 3" long
   Flowers: prolific, ⅛" wide, in axillary panicles
   Fruit: bright red spherical drupe, ⅛" diameter
   Schinus terebinthifolius (148)
   h. Leaflets: 5 to 9, opposite pairs, oblong to ovate, minutely scalloped margins, dotted with small translucent glandular windows
   Flowers: 5 greenish-white petals in terminal panicles
   Fruit: small ovoid brown drupes with single black seed
   Zanthoxylum flavum (170)

2. Leaves Bipinnate
   a. Leaves: 4 to 8 pairs pinnae, leaflets ½ to 1″ long, slender, elliptic
      Flowers: fluffy globular clusters, 1 to 1½″ diameter, axillary spikes
      Fruit: flat, brown pod, 6 to 8″ long
                                    Leucaena leucocephala (106)
   b. Leaves: 4 to 8 pinnae, leaflets ½″ long, oblong
      Flowers: small, greenish-white, fuzzy globular clusters, ½ to 1″ diameter
      Fruit: thin flat pod, 3 to 6″ long, reddish-brown, 8 to 10 brown seeds   *Lysiloma latisiliquum* (108)
C. Flowers—Red
  1. Leaves Pinnate
    a. Leaflets: 5 to 9 oval or elliptic-ovate 1½ to 4″ long, entire, wavy margins
       Flowers: pea-like, pink and white, 1″ wide
       Fruit: brown flat pods, 3 to 4″ long, 1″ wide, ruffled papery margins   *Piscidia piscipula* (123)
  2. Leaves Bipinnate
    a. Flowers: small, fragrant, pink globular heads, ¾ to 1¼″ wide
       Fruit: brown seed pods, 2 to 4″ long, opens when ripe, twisted, coiled, black seed, red arils
       Leaves: pair of bifoliate pinnae, 2 pairs of leaflets, broad oblong, notched or pointed tip, veins prominent, ½ to 2½″ long
       (1) Leaflet stems longer than leafstems
          Stipular spines rare
                  Pithecellobium guadalupense (126)
       (2) Leafstems longer than leaflet stems
          Stipular spines common
                              P. ungis-cati (127)
  3. Leaves Trifoliolate
    a. Leaflets: three arrowhead-shaped, 1 to 3″ long
       Flowers: corolla scarlet, erect spike, wing and keel petals 1 to 1½″ long
       Fruit: 3 to 6″ long legume, blackish, constricted between bright red seeds
                                  Erythrina herbacea (78)

D. Flowers—Green
1. Leaves Pinnate
   a. Leaflets: 3 to 9, ovate to oblong, pointed 2 to 3″ long, aromatic
      Flowers: small, many-flowered panicles, fragrant
      Fruit: red, spherical, ¼ to ½″ diameter
      Bursera simaruba (26)
   b. Leaflets: 7 to 9, ovate to obovate, ½ to 1½″ long, margins notched, leafstem winged
      Flowers: small, axillary clusters
      Fruit: brown ovoid, small, black seeds
      Zanthoxylum fagara (169)
2. Leaves Trifoliate
   a. Leaflets: 3 wedge-shaped, slightly notched at apex, close parallel veins
      Flowers: ⅛″ wide, axillary clusters
      Fruit: black oval drupes, ⅜″ diameter
      Hypelate trifoliata (98)

IX. LEAVES—COMPOUND—OPPOSITE
A. Flowers—Yellow
1. Leaves Pinnate
   a. Leaflets: 5 to 13, lanceolate to elliptic serrated margins, 3 to 5″ long, pointed
      Flowers: tubular, 2″ long, flared 1″ across terminal clusters
      Fruit: bean-like capsule to 8″ long
      Tecoma stans (160)
2. Leaves Bipinnate
   a. Leaves: 4 or more pairs of pinnae, leaflets ovate, 1 to 2″ long, 3 to 5 pairs
      Flowers: 5 petals, ½″ across, axillary racemes
      Fruit: oval-elliptic, flattened, spiny, reddish brown pods, seeds oval, ¾″ diameter
      (1) Gray seeds   *Caesalpinia bonduc* (28)
      (2) Yellow seeds   *C. major* (29)
   b. Leaves: 4 or more pairs of pinnae
      Leaflets: 5 to 7 pairs, oblong, apex slightly notched, ½″ long
      Flowers: flared curled back petals, ½″ across
      Fruit: pea-shaped pod, dark brown, 1 to 1½″ long, pointed at each end
      Spines: paired, stiff ¼″, axillary
      C. puciflora (30)

B. Flowers—White
   1. Leaves Pinnate
      a. Leaflets: 3 to 5, drooping, ovate-lanceolate, 1 to 2½" long, entire or faintly scalloped margins, long tapered apex
         Flowers: 4 petals, terminal clusters, fragrant
         Fruit: globular, ½" drupe, black or dark purple
         Amyris elemifera (11)
C. Flowers—Blue
   1. Leaves Pinnate
      a. Leaflets: 3 to 4 pairs, obovate, subsessile, 1" long, small sharp point at apex
         Flowers: 5 petals, star-like, 1" across
         Fruit: obovoid, 5-angled, ¾" across, splits open, black seeds, red arils   *Guaiacum sanctum* (89)

INDEX TO FAMILY GROUPS

The numbers in parenthesis beside each plant species name indicate the page number in the text where further information is found.)

AGAVACEAE	AGAVE FAMILY
Agave decipiens (7)	False Sisal
Agave sisalana (8)	Sisal Hemp
Yucca aloifolia (168)	Spanish Bayonet
ANACARDIACEAE	CASHEW FAMILY
Metopium toxiferum (114)	Poisonwood
Schinus terebinthifolia (148)	Brazilian Pepper
ANNONACEAE	CUSTARD APPLE FAMILY
Annona glabra (12)	Pond Apple
APOCYNACEAE	OLEANDER FAMILY
Vallesia antillana (166)	Pearl-berry
AQUIFOLIACEAE	HOLLY FAMILY
Ilex cassine (99)	Dahoon Holly
ARECACEAE	PALM FAMILY
Acoelorrhaphe wrightii (6)	Paurotis Palm
Coccothrinax argentata (57)	Silver Palm
Cocos nucifera (58)	Coconut Palm
Pseudophoenix sargentii (130)	Buccaneer Palm
Roystonea elata (142)	Royal Palm
Sabal palmetto (143)	Cabbage Palm
Serenoa repens (150)	Saw Palmetto
Thrinax morrisii (162)	Key Thatch Palm
Thrinax radiata (163)	Florida Thatch Palm
ASTERACEAE	ASTER FAMILY
Baccharis angustifolia (17)	Saltbush
Baccharis halimifolia (18)	Groundsel Tree
Borrichia arborescens (19)	Sea Ox-eye
Borrichia frutescens (20)	Sea Daisy
Pluchea odorata (128)	Marsh Fleabane
Pluchea symphytifolia (129)	Bushy Fleabane
AVICENNIACEAE	BLACK MANGROVE FAMILY
Avicennia germinans (16)	Black Mangrove
BIGNONIACEAE	BIGNONIA FAMILY
Amphitecna latifolia (10)	Black Calabash
Crescentia cujete (66)	Calabash Tree
Tecoma stans (160)	Yellow Elder

BORAGINACEAE
   Bourreria cassinifolia (21)
   Bourreria ovata (22)
   Bourreria radula (23)
   Cordia globosa (64)
   Cordia sebestena (65)
   Mallotonia gnaphalodes (109)
BURSERACEAE
   Bursera simaruba (26)
CACTACEAE
   Cereus gracilis (44)
   Cereus pentagonis (45)
   Cereus robinii (46)
   Opuntia spinosissima (119)
   Opuntia stricta (120)
CANELLACEAE
   Canella winterana (34)
CAPPARACEAE
   Capparis cynophallophora (35)
   Capparis flexuosa (36)
CARICACEAE
   Carica papaya (37)
CASUARINACEAE
   Casuarina equisetifolia (41)
   Casuarina glauca (42)
CELASTRACEAE
   Crossopetalum ilicifolium (67)
   Crossopetalum rhacoma (68)
   Gyminda latifolia (93)
   Maytenus phyllanthoides (113)
   Schaefferia frutescens (147)
CHRYSOBALANACEAE
   Chrysobalanus icaco (49)
CLUSIACEAE
   Clusia rosea (54)
COMBRETACEAE
   Conocarpus erectus (62)
   Conocarpus erectus sericeus (63)
   Laguncularia racemosa (102)

BORAGE FAMILY
   Smooth Strongbark
   Strongbark
   Rough Strongbark
   Bloodberry
   Geiger Tree
   Sea Lavender
GUMBO-LIMBO FAMILY
   Gumbo Limbo
CACTUS FAMILY
   Prickly Apple
   Dildo Cactus
   Tree Cactus
   Semaphore Cactus
   Prickly Pear
WILD-CINNAMON FAMILY
   Cinnamon Bark
CAPER FAMILY
   Jamaica Caper
   Limber Caper
PAPAYA FAMILY
   Papaya
AUSTRALIAN PINE FAMILY
   Australian Pine
   Brazilian Oak
BITTERSWEET FAMILY
   Christmas-berry
   Rhacoma
   False Boxwood
   Mayten
   Florida Boxwood
COCO-PLUM FAMILY
   Coco-plum
CLUSIA FAMILY
   Autograph Tree
COMBRETUM FAMILY
   Buttonwood
   Silver Buttonwood

   White Mangrove

EUPHORBIACEAE | RUBBER FAMILY
   Ateramnus lucidus (15) | Crabwood
   Croton linearis (69) | Wild Croton
   Drypetes diversifolia (73) | Milk Bark
   Drypetes lateriflora (74) | Guiana Plum
   Hippomane mancinella (97) | Manchineel
   Ricinus communis (140) | Caster Bean
   Rivina humilis (141) | Bloodberry
   Savia bahamensis (145) | Maiden Bush

FABACEAE — PEA FAMILY
- *Acacia choriophylla* (2) — Cinnecord
- *Acacia farnesiana* (3) — Sweet Acacia
- *Acacia macracantha* (4) — Long Spined Acacia
- *Acacia pinetorum* (5) — Pine Acacia
- *Albizia lebbeck* (9) — Woman's Tongue
- *Caesalpinia bonduc* (28) — Gray Nicker Bean
- *Caesalpinia major* (29) — Yellow Nicker Bean
- *Caesalpinia pauciflora* (30) — Caesalpinia
- *Cassia chapmanii* (39) — Bahama Senna
- *Cassia keynesis* (40) — Key Cassia
- *Dalbergia brownei* (71) — Coin Vine
- *Erythrina herbacea* (78) — Coral Bean
- *Leucaena leucocephala* (106) — Lead Tree
- *Lysiloma latisiliquum* (108) — Wild Tamarind
- *Piscidia piscipula* (123) — Jamaica Dogwood
- *Pithecellobium guadalupense* (126) — Blackbead
- *Pithecellobium ungis-cati* (127) — Catclaw
- *Quercus virginiana* (136) — Live Oak
- *Sophora tomentosa* (155) — Necklace-pod

GOODENIACEAE — GOODENIA FAMILY
- *Scaevola plumieri* (146) — Ink-berry

LAURACEAE — LAUREL FAMILY
- *Nectandra coriaceae* (118) — Lancewood
- *Persea borbonea* (121) — Red Bay

MALPIGHIACEAE — MALPIGHIA FAMILY
- *Byrsonima lucida* (27) — Locust-berry

MALVACEAE — MALLOW FAMILY
- *Abutilon permolle* (1) — Indian Mallow
- *Cienfuegosia yucataniensis* (51)
- *Gossypium hirsutum* (88) — Wild Cotton
- *Hibiscus poeppigii* (95) — Wild Hibiscus
- *Hibiscus tiliaceus* (96) — Mahoe
- *Thespesia populnea* (161) — Portia Tree

MELIACEAE   MAHOGANY FAMILY
Swietenia mahagoni (159)   Mahogany
MORACEAE   MULBERRY FAMILY
Ficus aurea (85)   Strangler Fig
Ficus citrifolia (86)   Shortleaf Fig
MYRICACEAE   BAYBERRY FAMILY
Myrica cerifera (116)   Wax-myrtle
MYRSINACEAE   MYRSINE FAMILY
Ardisia escallonioides (13)   Marlberry
Myrsine floridana (117)   Myrsine
MYRTACEAE   MYRTLE FAMILY
Calyptranthes pallens (32)   Pale Lid-flower
Calyptranthes zuzygium (33)   Myrtle-of-the-River
Eugenia axillaris (79)   White Stopper
Eugenia confusa (80)   Red-berry Stopper
Eugenia foetida (81)   Spanish Stopper
Eugenia rhombea (82)   Red Stopper
Psidium guajava (131)   Guava
Psidium longipes (132)   Long-stalked Stopper
NYCTAGINACEAE   FOUR-O'CLOCK FAMILY
Guapira discolor (90)   Blolly
Pisonia aculeata (124)   Cockspur
Pisonia rotundata (125)   Pisonia
OLACACEAE   SCHOEPFIA FAMILY
Schoepfia chrysophylloides (149)   Graytwig
Ximenia americana (167)   Tallowwood
OLEACEAE   OLIVE FAMILY
Forestiera segregata (87)   Florida Privet
PINACEAE   PINE FAMILY
Pinus elliottii (122)   Slash Pine
POACEAE   GRASS FAMILY
Lasiacis divaricata (105)   Wild Bamboo
POLYGONACEAE   BUCKWHEAT FAMILY
Coccoloba diversifolia (55)   Pigeon Plum
Coccoloba uvifera (56)   Sea-grape
RHAMNACEAE   BUCKTHORN FAMILY
Colubrina arborescens (59)   Wild Coffee
Colubrina asiatica (60)   Latherleaf
Colubrina elliptica (61)   Soldierwood
Krugiodendron ferreum (101)   Black Ironwood
Reynosia septentrionalis (138)   Darling Plum

RHIZOPHORACEAE — RED MANGROVE FAMILY
  Rhizophora mangle (139) — Red Mangrove
RUBIACEAE — MADDER FAMILY
  Casasia clusiifolia (38) — Seven-year Apple
  Catesbaea parviflora (43) — Small-flowered Lily-thorn
  Chiococca alba (47) — Snowberry
  Chiococca parvifolia (48) — Snowberry
  Erithalis fruticosa (76) — Black Torch
  Ernodea littoralis (77) — Golden Creeper
  Exostema caribaeum (83) — Princewood
  Guettarda elliptica (91) — Velvet-seed
  Guettarda scabra (92) — Rough Velvetseed
  Hamelia patens (94) — Scarlet Bush
  Morinda royoc (115) — Yellow Root
  Psychotria ligustrifolia (133) — Wild Coffee
  Psychotria nervosa (134) — Wild Coffee
  Psychotria punctata (135) — Wild Coffee
  Randia aculeata (137) — Randia
  Strumpfia maritima (157) — Pride-of-Big-Pine
RUTACEAE — RUE FAMILY
  Amyris elemifera (11) — Torchwood
  Citrus aurantifolia (53) — Key Lime
  Zanthoxylum fagara (169) — Wild Lime
  Zanthoxylum flavum (170) — Satinwood
SAPINDACEAE — SOAPBERRY FAMILY
  Cupania glabra (70) — Florida Cupania
  Dodonaea viscosa (72) — Florida Hop Bush
  Exothea paniculata (84) — Inkwood
  Hypelate trifoliata (98) — White Ironwood
  Sapindus saponaria (144) — Wingleaf Soapberry
SAPOTACEAE — SAPODILLA FAMILY
  Bumelia celastrina (24) — Saffron Plum
  Bumelia salicifolia (25) — Willow Bustic
  Chrysophyllum oliviforme (50) — Satin Leaf
  Manilkara bahamensis (110) — Wild Dilly
  Manilkara zapota (111) — Sapodilla
  Mastichodendron foetidissimum (112) — Mastic
SIMAROUBACEAE — BITTERBARK FAMILY
  Simarouba glauca (151) — Paradise Tree

SOLANACEAE
Lycium carolinianum (107)
Solanum bahamense (152)
Solanum donianum (153)
Solanum erianthum (154)
SURIANACEAE
Suriana maritima (158)
THEOPHRASTACEAE
Jacquinia keyensis (100)
ULMACEAE
Trema lamarckianum (164)
Trema micranthum (165)
VERBENACEAE
Callicarpa americana (31)
Citharexylum fruticosum (52)
Duranta repens (75)
Lantana camara (103)
Lantana involucrata (104)
Stachytarpheta jamaicensis (156)
ZYGOPHYLLACEAE
Guaiacum sanctum (89)

NIGHTSHADE FAMILY
Christmas Berry
Bahama Nightshade
Potato Tree
Mullein Nightshade
BAY CEDAR FAMILY
Bay Cedar
JOEWOOD FAMILY
Joewood
ELM FAMILY
West Indies Trema
Florida Trema
VERBENA FAMILY
Beauty Berry
Fiddlewood
Golden Dewdrop
Lantana
Wild Lantana
Blue Porterweed

LIGNUM-VITAE FAMILY
Lignum Vitae

INDEX TO SPECIAL GROUP

MANGROVE COMMUNITY
Avicennia germinans (16)
Rhizophora mangle (139)
Laguncularia racemosa (102)
Conocarpus erectus var. *sericeus* (63)

Black Mangrove
Red Mangrove
White Mangrove
Silver Buttonwood

KEYS' TREES AND SHRUBS DESIGNATED STATUS

SCIENTIFIC NAME	COMMON NAME	STATUS*	PAGE
1. *Acacia choriophylla*	Tamarandillo	Endangered	2
2. *Acoelorrhaphe wrightii*	Paurotis Palm	Threatened	6
3. *Argythamnia blodgettii*	Blodgett's Wild-cherry	Threatened	14
4. *Avicennia germinans*	Black Mangrove	Special Concern	16
5. *Cassia keyensis*	Big Pine Partridge Pea	Endangered	40
6. *Catesbaea parviflora*	Small-flowered Lily-thorn	Endangered	43
7. *Cereus gracilis*	Prickly-apple	Endangered	44
8. *Cereus pentagonus*	Barbed Wire Cactus	Threatened	45
9. *Cereus robinii*	Tree Cactus	Endangered	46
10. *Chrysophyllum oliviforme*	Satinleaf	Threatened	50
11. *Clusia rosea*	Balsam-apple	Endangered	54
12. *Coccothrinax argentata*	Silver Palm	Endangered	57
13. *Cordia sebestena*	Geiger Tree	Threatened	65
14. *Cupania glabra*	Cupania	Endangered	70
15. *Ernodea littoralis*	Beach Creeper	Threatened	77
16. *Eugenia confusa*	Red-berry Stopper	Threatened	80
17. *Eugenia rhombea*	Red Stopper	Endangered	82
18. *Gossypium hirsutum*	Wild Cotton	Endangered	88
19. *Guaiacum sanctum*	Lignum Vitae	Endangered	89
20. *Hippomane mancinella*	Manchineel	Threatened	97
21. *Hypelate trifoliata*	Inkwood	Threatened	98
22. *Mallotonia gnaphalodes*	Sea Lavender	Threatened	109
23. *Opuntia spinosissima*	Semaphore Cactus	Threatened	119
24. *Opuntia stricta*	Prickly Pear	Threatened	120
25. *Pseudophoenix sargentii*	Buccaneer Palm	Endangered	130

26. *Rhizophora mangle*	Red Mangrove	Special Concern	139
27. *Roystonea elata*	Florida Royal Palm	Endangered	142
28. *Scaevola plumieri*	Scaevola	Endangered	146
29. *Strumpfia maritima*	Pride-of-Big-Pine	Endangered	157
30. *Suriana maritima*	Bay Cedar	Endangered	158
31. *Swietenia mahogoni*	Mahogany	Threatened	159
32. *Thrinax morrisii*	Brittle Thatch Palm	Threatened	162
33. *Thrinax radiata*	Florida Thatch Palm	Threatened	163
34. *Zanthoxylum flavum*	Yellow-heart	Endangered	170

*Definition of Designated Status

Endangered. Species in imminent danger of extinction or extirpation and whose survival is unlikely if the causal factors presently at work continue operating.

Threatened. Species believed likely to move into the Endangered category in the near future if the causal factors now at work continue operating.

Rare. Species with small populations in the state which, though not presently Endangered or Threatened as defined above, are potentially at risk. Included are species that may be localized within a restricted geographical region or habitat or thinly scattered over a more extensive range.

Species of Special Concern. A species that does not clearly fit into the Endangered, Threatened or Rare categories yet which, for certain reasons warrants special attention.

*From *Rare and Endangered Biota of Florida* Volume 5 *Plants,* edited by Daniel B. Ward, University of Florida Presses (1978).

INDEX TO SCIENTIFIC AND COMMON NAMES

Since the purpose of this book is primarily to help people identify native and naturalized plants in the Keys, an attempt was made to create an index that would be a usuable tool in the identification process. The index is a consolidated alphabetical list of the scientific names used in this book, the scientific synonyms used in the past and the common or local names by which people from different parts of the world recognize the plants.

The numbers following the names indicate the page number in the text where the plant is found.

INDEX TO SCIENTIFIC AND COMMON NAMES

Abutilon permolle, 1
Acacia choriophylla, 2
Acacia farnesiana, 3
Acacia macracantha, 4
Acacia pinetorum, 5
Acanthocereus floridanus, 45
Achras emarginata, 110
Achras zapota, 111
Acoelorrhaphe wrightii, 6
Agave decipiens, 7
Agave sisalana, 8
Albizia lebbeck, 9
Alligator-apple, 12
Aloe Yucca, 168
American Beautyberry, 31
American Upland Cotton, 88
Amerimnon brownei, 71
Amphitecna latifolia, 10
Amyris elemifera, 11
Amyris maritima, 11
Annona glabra, 12
Annona palustris, 12
Antswood, 24
Ardisia escallonioides, 13
Argythamnia blodgettii, 14
Ateramnus lucidus, 15
Australian Pine, 41
Autograph Tree, 54
Avicennia germinans, 16
Avicennia nitida, 16
Baccharis angustifolia, 17
Baccharis halimifolia, 18
Bahama Lysiloma, 108
Bahama Nightshade, 152
Bahama Senna, 39
Bahama Strongbark, 22
Bahaman Wild Coffee, 133
Balsam Apple, 54
Barbasco, 100
Barbed-wire Cactus, 45
Bay Cedar, 158
Bay Lavender, 109
Bay-leaved Caper Tree, 36
Beach-berry, 146

Beach-creeper, 77
Beach Heliotrope, 109
Beautyberry, 31
Beefwood, 41, 90
Big Pine Partridge Pea, 40
Biscayne Palm, 57
Bitterwood, 151
Black Calabash, 10
Black Ironwood, 101
Black Mangrove, 16
Black Torch, 76
Black Willow, 35
Blackbead, 126
Blodgett's Wild-mercury, 14
Blolly, 90
Bloodberry, 64, 141
Blue Porterweed, 156
Bontia germinans, 16
Borrichia arborescens, 19
Borrichia frutescens, 20
Bourreria cassinifolia, 21
Bourreria ovata, 22
Bourreria radula, 23
Bourreria revoluta, 23
Bourreria succulenta, 23
Box-brier, 137
Box-leaf Stopper, 81
Boxwood, 147
Brazilian Oak, 42
Brazilian Pepper, 148
Brittle Thatch Palm, 162
Buccaneer Palm, 130
Bull-bay, 121
Bumelia angustifolia, 24
Bumelia celastrina, 24
Bumelia salicifolia, 25
Bursera simaruba, 26
Bushy Fleabane, 129
Bustic, 25
Butter-bough, 84
Button-mangrove, 62
Buttonwood, 62
Byrsonima cuneata, 27
Byrsonima lucida, 27

INDEX TO SCIENTIFIC AND COMMON NAMES

Cabbage Palm, 143
Caesalpinia bonduc, 28, 29
Caesalpinia crista, 28
Caesalpinia major, 29
Caesalpinia pauciflora, 30
Calabash Tree, 66
Callicarpa americana, 31
Calyptranthes pallens, 32
Calyptranthes zuzygium, 33
Canella alba, 34
Canella winterana, 34
Canker-berry, 152
Caper Tree, 35
Capparis cynophallophora, 35
Capparis flexuosa, 36
Capparis jamaicensis, 35
Cardinal Spear, 78
Caribbean Princewood, 83
Carica papaya, 37
Carolina Palm, 143
Casasia clusiifolia, 38
Cassada, 25
Cassena, 99
Cassia bahamensis, 39
Cassia chapmanii, 39
Cassia keyensis, 40
Cassie, 3
Castor Bean, 140
Castor Oil Plant, 140
Casuarina equisetifolia, 41
Casuarina glauca, 42
Casuarina lepidophloia, 42
Casuarina littorea, 41
Cat's Claw, 127
Catclaw, 127
Catclaw Blackbead, 127
Catesbaea parviflora, 43
Century Plant, 7, 8
Cereus gracilis, 44
Cereus pentagonus, 45
Cereus robinii, 46
Cerothamnus ceriferus, 116
Chamaecrista keyensis, 40
Cheese Shrub, 115

Cherokee Bean, 78
Cherry, 13
Chicle-gum Tree, 111
Chiococca alba, 47
Chiococca parvifolia, 48
Chiococca pinetorum, 48
Christmas Berry, 107, 148
Christmas-berry, 67
Chrysobalanus icaco, 49
Chrysobalanus interior, 49
Chrysobalanus pellocarpus, 49
Chrysophyllum oliviforme, 50
Cienfuegosia yucataniensis, 51
Cinnamon Bark, 34
Cinnecord, 2
Citharexylum fruticosum, 52
Citharexylum villosum, 52
Citrus aurantifolia, 53
Citrus lima, 53
Clusia rosea, 54
Coccoloba diversifolia, 55
Coccoloba floridana, 55
Coccoloba laurifolia, 55
Coccoloba uvifera, 56
Coccothrinax argentata, 57
Coccothrinax garberi, 57
Coccothrinax jucunda, 57
Cockspur, 124
Coco-plum, 49
Coconut Palm, 58
Cocos nucifera, 58
Coffee Colubrina, 59
Coin Vine, 71
Colubrina arborescens, 59
Colubrina asiatica, 60
Colubrina colubrina, 59
Colubrina elliptica, 61
Colubrina reclinata, 61
Column Cactus, 46
Common Cotton, 88
Conocarpus erectus, 62
Conocarpus erectus var. *sericeus,* 63
Consolea corallicola, 119

INDEX TO SCIENTIFIC AND COMMON NAMES

Copey, 54
Coral Bean, 78
Coral Sumac, 114
Cordia globosa, 64
Cordia sebestena, 65
Cork Tree, 161
Corkwood, 12
Cotton, 88
Crabwood, 15
Crescentia cujete, 66
Crossopetalum ilicifolium, 67
Crossopetalum rhacoma, 68
Croton linearis, 69
Cudjoe-wood, 100
Cupania glabra, 70
Custard-apple, 12
Dahoon Holly, 99
Dalbergia amerimnon, 71
Dalbergia brownei, 71
Darling Plum, 138
Devil's Claws, 124
Dildo Cactus, 45
Dillen's Prickly Pear, 120
Dipholis salicifolia, 25
Doctor Gum, 114
Ditaxis blodgettii, 14
Dodonaea jamaicensis, 72
Dodonaea microcarya, 72
Dodonaea viscosa, 72
Dove Plum, 55
Downward Plum, 24
Drypetes diversifolia, 73
Drypetes keyensis, 73
Drypetes lateriflora, 74
Dune Lily-thorn, 43
Duranta repens, 75
Eastern Baccharis, 18
Eastern Coral Bean, 78
Elaphrium simaruba, 26
Elliptic-leaf Velvetseed, 91
Enallagma latifolia, 10
Erithalis fruticosa, 76
Ernodea littoralis, 77
Erythrina arborea, 78

Erythrina herbacea, 78
Eugenia anthera, 82
Eugenia axillaris, 79
Eugenia buxifolia, 81
Eugenia confusa, 80
Eugenia foetida, 81
Eugenia monticola, 79
Eugenia myrtoides, 81
Eugenia procera, 82
Eugenia rhombea, 82
Everglades Cabbage Palm, 6
Everglades Velvetseed, 91
Exostema caribaeum, 83
Exothea paniculata, 84
Fagara flava, 170
Fagara pterota, 169
False Boxwood, 93
False Mastic, 112
False Phoenix, 130
False Rosewood, 161
False Sisal, 7
False Willow, 17
Fever Nut, 28
Ficus aurea, 85
Ficus brevifolia, 86
Ficus citrifolia, 86
Ficus laevigata, 86
Ficus populnea, 86
Fiddlewood, 52
Firebush, 94
Fish-poison Tree, 123
Fishfuddle Tree, 123
Florida Boxwood, 147
Florida Crossopetalum, 68
Florida Cupania, 70
Florida Fiddlewood, 52
Florida Forestiera, 87
Florida Holly, 148
Florida Hop Bush, 72
Florida Mayten, 113
Florida Nectandra, 118
Florida Poisontree, 114
Florida Privet, 87
Florida Royal Palm, 142

INDEX TO SCIENTIFIC AND COMMON NAMES

Florida Silverpalm, 57
Florida Strangler Fig, 85
Florida Thatch Palm, 163
Florida Trema, 165
Forestiera segregata, 87
French Mulberry, 31
Frijollo, 2
Geiger Tree, 65
Genipa clusiifolia, 38
Geranium Tree, 65
Golden-creeper, 77
Golden Dewdrop, 75
Golden Fig, 85
Gossypium herbaceum, 88
Gossypium hirsutum, 88
Gossypium punctatum, 88
Granny-bush, 69
Gray Nicker Bean, 28
Graytwig, 149
Ground-holly, 67
Groundsel Tree, 18
Guadeloupe Blackbead, 126
Guaiacum guatemalense, 89
Guaiacum sanctum, 89
Guapira discolor, 90
Guapira longifolia, 90
Guava, 131
Guettarda elliptica, 91
Guettarda scabra, 92
Guiana Plum, 74
Guilandina bonduc, 28, 29
Guilandina crista, 28
Gulf Graytwig, 149
Gum-elemi, 26
Gumbo-limbo, 26
Gurgeon Stopper, 81
Gutta-percha Mayten, 113
Gyminda grisebachii, 93
Gyminda latifolia, 93
Gymnanthes lucida, 15
Hairy Sophora, 155
Hamelia erecta, 94
Hamelia patens, 94
Harrisia simpsonii, 44

Hibiscus pilosus, 95
Hibiscus poeppigii, 95
Hibiscus tiliaceus, 96
Hippomane mancinella, 97
Hog Gum, 114
Hog Palm, 130
Hog Plum, 167
Hold-back, 28
Holywood, 89
Honey Mangrove, 16
Hop Bush, 72
Horsetail Tree, 41
Huisache, 3
Hypelate trifoliata, 98
Icaco Coco-plum, 49
Icacorea paniculata, 13
Ichthyomethia piscipula, 123
Ilex cassine, 99
Indian Mallow, 1
Indigo-berry, 137
Ink-berry, 146
Inkwood, 84, 98
Ironwood, 84
Jacquinia keyensis, 100
Jamaica Caper, 35
Jamaica Dogwood, 123
Jamaica Nectandra, 118
Jamaica Thatch Palm, 163
Joewood, 100
Jumbie Bean, 106
Jungle Plum, 112
Key Byrsonima, 27
Key Cassia, 40
Key Lime, 53
Key Palm, 162
KeyThatch Palm, 162
Krugiodendron ferreum, 101
Laguncularia racemosa, 102
Lancewood, 118
Lantana aculeata, 103
Lantana camara, 103
Lantana involucrata, 104
Lantana odorata, 104
Lasiacis divaricata, 105

INDEX TO SCIENTIFIC AND COMMON NAMES

Latherleaf, 60
Lead Tree, 106
Leadwood, 101
Lebbeck Tree, 9
Leucaena glauca, 106
Leucaena leucocephala, 106
Lignum Vitae, 89
Limber Caper, 36
Lime Prickly-ash, 169
Little Strongback, 21
Live Oak, 136
Locust-berry, 27
Long-spine Acacia, 4
Long-stalked Stopper, 132
Longleaf Blolly, 90
Lycium carolinianum, 107
Lysiloma bahamensis, 108
Lysiloma latisiliquum, 108
Madeira Palm, 6
Mahoe, 96
Mahogany, 159
Maiden Bush, 145
Mallotonia gnaphalodes, 109
Manchineel, 97
Manilkara bahamensis, 110
Manilkara jaimiqui, 110
Manilkara zapota, 111
Manilkara zapotilla, 111
Manzanillo, 97
Marbleberry, 13
Marlberry, 13
Marsh Fleabane, 128
Mastic, 112
Mastichodendron foetidissimum, 112
Mayten, 113
Maytenus phyllanthoides, 113
Melon-tree, 37
Metopium toxiferum, 114
Milkbark, 73
Mimosa glauca, 106
Mimosa lebbeck, 9
Mimosa unguis-cati, 127
Mimusops emarginata, 110

Monkey Apple, 54
Morinda royoc, 115
Mother-in-Law's Tongue, 9
Mouse's Pineapple, 115
Mullein Nightshade, 154
Myrica cerifera, 116
Myrsine floridana, 117
Myrsine guianensis, 117
Myrtle-of-the-River, 33
Naked-wood, 61
Naseberry, 111
Necklace-pod, 155
Nectandra coriacea, 118
Ocotea catesbyana, 118
Opopanax, 3
Opuntia dillenii, 120
Opuntia keyensis, 120
Opuntia spinosissima, 119
Opuntia stricta, 120
Opuntia zebrina, 120
Oval-leaf Strongbark, 22
Oysterwood, 15
Pale Lidflower, 32
Palmetto Palm, 143
Papaya, 37
Paradise Tree, 151
Pariti tiliaceum, 96
Paurotis Palm, 6
Paurotis wrightii, 6
Pearl-berry, 166
Pepper Tree, 148
Persea borbonia, 121
Persea littoralis, 121
Persea palustris, 121
Pigeon-berry, 75
Pigeon Plum, 55
Pine acacia, 5
Pineland Croton, 69
Pinus caribaea, 122
Pinus elliottii, 122
Piscidia erythrina, 123
Piscidia piscipula, 123
Pisonia aculeata, 124
Pisonia discolor, 90

INDEX TO SCIENTIFIC AND COMMON NAMES

Pisonia rotundata, 125
Pitch Apple, 54
Pithecellobium guadalupense, 126
Pithecellobium keyense, 126
Pithecellobium unguis-cati, 127
Pluchea odorata, 128
Pluchea symphytifolia, 129
Poison-guava, 97
Poisonwood, 114
Pond Apple, 12
Popinac, 3
Poponax macracantha, 4
Portia Tree, 161
Potato Tree, 153, 154
Prickly Apple, 44
Prickly Pear, 120
Pride-of-Big-Pine, 157
Princewood, 83
Pseudophoenix sargentii, 130
Psidium guajava, 131
Psidium longipes, 132
Psychotria bacteriophylla, 135
Psychotria bahamensis, 133
Psychotria ligustrifolia, 133
Psychotria nervosa, 134
Psychotria punctata, 135
Psychotria undata, 134
Pull-and-Hold-Back, 124
Purge Nut, 167
Quail-berry, 67
Quercus virginiana, 136
Ram's Horn, 126
Randia aculeata, 137
Rapanea, 117
Rapanea guianensis, 117
Rapanea punctata, 117
Rat's Tail, 156
Red Bay, 121
Red-bay Persea, 121
Red Ironwood, 138
Red Mangrove, 139
Red Stopper, 82

Red-berry Eugenia, 80
Red-berry Stopper, 80
Reynosia septentrionalis, 138
Rhacoma, 68
Rhacoma crossopetalum, 68
Rhacoma ilicifolia, 67
Rhamnium ferreum, 101
Rhizophora mangle, 139
Rhus terebinthifolia, 148
Ricinus communis, 140
Rivina humilis, 141
Rock Pine, 122
Rouge Plant, 141
Rough Strongback, 23
Rough Strongbark, 23
Rough Velvetseed, 92
Roughbark Lignum Vitae, 89
Roughleaf Velvetseed, 92
Royal Palm, 142
Roystonea elata, 142
Roystonea floridana, 142
Roystonea regia, 142
Sabal jamesiana, 143
Sabal palmetto, 143
Saffron Plum, 24
Saltbush, 16, 17, 18
Sapindus saponaria, 144
Sapodilla, 111
Sargent's Cherry Palm, 130
Satin Leaf, 50
Satinwood, 170
Savia bahamensis, 145
Saw Cabbage Palm, 6
Saw Palmetto, 150
Scaevola plumieri, 146
Schaefferia frutescens, 147
Scalybark Beefwood, 42
Scarlet-bush, 94
Scarlet Cordia, 65
Schinus terebinthifolia, 148
Schoepfia americana, 149
Schoepfia chrysophylloides, 149
Schoepfia schreberi, 149

INDEX TO SCIENTIFIC AND COMMON NAMES

Scotch Attorney, 54
Sea Bean, 28
Sea Daisy, 20
Sea-grape, 56
Sea Hibiscus, 96
Sea Lavender, 109
Sea Myrtle, 18, 100
Sea Ox-eye, 19
Seamberry Palm, 57
Seaside-grape, 56
Seaside Mahoe, 161
Sebestena sebestena, 65
Semaphore Cactus, 119
Serenoa repens, 150
Seven-year Apple, 38
She-oak, 41
Shore-grape, 56
Shortleaf Fig, 86
Shrub Verbena, 103
Signature Tree, 54
Silver Bush, 155
Silver Buttonwood, 63
Silver-leaved Buttonwood, 63
Silver Palm, 57
Silver-saw Palmetto, 6
Silverling, 18
Simarouba glauca, 151
Siris Tree, 9
Sisal, 8
Sisal Agave, 8
Sisal Hemp, 8
Sky-flower, 75
Slash Pine, 122
Small-flowered Lily-thorn, 43
Smooth Strongbark, 21
Snakebark, 59
Snakeroot, 47
Snakeweed, 78
Snowbank, 157
Snowberry, 47, 48
Soapberry, 144
Solanum bahamense, 152
Solanum donianum, 153
Solanum erianthum, 154

Solanum verbascifolium, 154
Sophora tomentosa, 155
Soldierwood, 61
South Florida Slash Pine, 122
Southern Bayberry, 116
Southern Wax-myrtle, 116
Spanish Bayonet, 168
Spanish Dagger, 168
Spanish Stopper, 81
Spiceberry, 82
Spiceberry eugenia, 82
Spicewood, 32, 33
Stachytarpheta jamaicensis, 156
Stopper, 79, 80, 81, 82
Strangler Fig, 85
Strongback, 22
Strongbark, 22
Strumpfia maritima, 157
Suriana maritima, 158
Swamp-bay, 121
Swamp Red Bay, 121
Swamp She-oak, 42
Sweet Acacia, 3
Sweet-bay, 121
Swietenia mahagoni, 159
Tallowwood, 167
Tamala borbonea, 121
Tamala pubescens, 121
Tamarindillo, 2
Tear-shrub, 166
Tears-of-St. Peter, 47
Tecoma stans, 160
Thespesia populnea, 161
Thrinax floridana, 163
Thrinax keyensis, 162
Thrinax microcarpa, 162
Thrinax morrisii, 162
Thrinax parviflora, 163
Thrinax radiata, 163
Tisswood, 121
Torchwood, 11
Torrubia globosa, 90
Torrubia longifolia, 90

INDEX TO SCIENTIFIC AND COMMON NAMES

Torrubia rotundata, 125
Tourist Tree, 26
Tournefortia gnaphalodes, 109
Trailing Eugenia, 132
Tree Cactus, 46
Tree Hibiscus, 96
Tree-of-life, 89
Trema floridana, 165
Trema lamarckianum, 164
Trema micranthum, 165
Upland Cotton, 88
Vachellia farnesiana, 3
Vachellia peninsularis, 5
Valerianoides jamaicensis, 156
Vallesia antillana, 166
Varnish-leaf, 72
Varronia globosa, 64
Velvet-seed, 91
Vomitel, 65
Wait-a-bit Vine, 28
Wattle, 79
Wax-myrtle, 116
West Indian Birch, 26
West Indies False-box, 93
West Indies Mahogany, 159
West Indies Satinwood, 170
West Indies Trema, 164
White Buttonwood, 102
White Indigo-berry, 137
White Ironwood, 98
White Mangrove, 102
White Popinac, 106
White Stopper, 79
White Stopper Eugenia, 79
Whitewood, 73, 149
Wild Bamboo, 105
Wild Banyan, 86
Wild Cinnamon, 34
Wild Coffee, 59, 133, 134, 135
Wild Cotton, 88
Wild Croton, 69
Wild Dilly, 110
Wild Fig, 85, 86
Wild Hibiscus, 95
Wild Lantana, 104
Wild Lime, 169
Wild Olive, 87, 112
Wild Sage, 104
Wild Sapodilla, 110
Wild Tamarind, 108
Willow Bustic, 25
Wingleaf Soapberry, 144
Woman's Tongue, 9
Ximenia americana, 167
Yaupon, 99
Yellow Elder, 160
Yellow-heart, 170
Yellow Nicker, 29
Yellow Nicker-bean, 29
Yellow Root, 115
Yellow-sage, 103
Yellow-wood, 147, 170
Yucca aloifolia, 168
Zanthoxylum fagara, 169
Zanthoxylum flavum, 170

218

ABOUT THE AUTHOR

James Paul Scurlock became interested in plants and trees in a large undeveloped forest near his childhood home in Greensboro, North Carolina, where he was born on May 14, 1909. As an Eagle Scout and a nature-study counselor at a Boy Scout camp, he developed his interest. While pursuing his Bachelor's degree in Civil Engineering at the University of North Carolina at Chapel Hill, he supplemented his college work with courses in biology and botany.

His interest in the Florida Keys began with an undergraduate assignment to study the 150 miles and 37 bridges of the "railroad that went to sea," the railroad that later, in the hurricane of 1935, literally went to sea. In 1928, Scurlock and two friends rode the train to Key West and sailed to Cuba. Riding on the rear platform of the last car over island after island of at that time undisturbed hardwood hammocks, Scurlock vowed to return and try to save as much as possible for people yet to come.

Serving in World War II in the United States Army Signal Corps, his first assignment was to form the First Signal Photographic Company, where he studied photography with teachers such as Elliott Elisofen of *Life* magazine. Overseas more than three years, more than two in the combat zone, and part of six major campaigns, Scurlock took refuge when he could in studying and photographing plants throughout the British Isles, North Africa, France, Italy, and Egypt.

Scurlock returned to the Keys in 1972. He was a retired executive of the Bell Telephone Company and a Colonel in the U.S. Army Reserves. He had continued his study as naturalist and photographer in state and national parks from the east coast to the west; he perfected his method of shooting his subjects undisturbed, coping with problems of wind and low light. He focused his study on the plants of the tropics as he travelled with Fairchild Tropical Garden groups to botannical gardens in Central and South America, including Costa Rica, Venezuela, Columbia, Guatemala, Brazil, Trinidad and Tobago.

Recognizing the similarities—and differences—of the Keys' plants to other tropicals, he searched for, could not find, and so began to create a photographic record identifying the Keys' natives as they occur uniquely in the local field.

Scurlock is President of the Florida Keys Recreational and Conservation Council and Chairman of the Key West Garden Club's Conservation and Environmental Committee. He is former President of the Key West Art and Historical Society, former President of the Sugarloaf Shores Property Owners Association, Board Member Emeritus of the Florida Keys Land Trust, member of the Monroe County Highway Beautification Committee, and Board Member of the Florida Keys Chapter of the Audubon Society.

Scurlock has lived on Lower Sugarloaf Key since 1972, where he continues his work as a naturalist. To date he has over 27,000 slides and color negatives—numbered, filed, described, dated and located geographically. He gives slide-illustrated lectures to garden clubs and botanical societies, civic and governmental groups, public schools and youth camps. He has developed and maintains a collection of native and exotic plants, using simple, organic methods—he relies on rainfall and compost; he uses neither fertilizers nor pesticides. The Florida Keys Land and Sea Trust presented to him the 1990 Ross McKee Award for his contribution to the protection of the natural heritage of the Florida Keys.

NOTES

NOTES

NOTES

NOTES

NOTES

NOTES

NOTES

NOTES

NOTES

NOTES